NAME OF GOD

By Exodus 3.15
I AM is His name for ever
And I AM is His memorial for all generations

*4 SEPT 2016, BREAKING NEWS –
GLOBAL BIBLICAL SPIRITUAL FLOOD
STARTS, SEE PAGE 154*

TRUE GNOSIS

Peu de science èloigne de Dieu
Beaucoup the science ya raméne

> *A little knowledge keeps one from God*
> *True knowledge brings back to Him*

> *Siva the cosmic ruler has failed and 3nity = Babylon is fallen. Eternal Son was being matured and now He is personalized by Maha Avatar Lord Kalki – prior to this and as interim measure the 3nity created and managed the cosmic life. But they refused to hand over the charge and decided to fight a secret long thought of, and undeclared war against the Lord Kalki. Siva always misled the cosmos that he was the Absolute God and failed in the final show down – thus Truth won Justice Prevails Radiates and Spreads out*

SWASTI

Root meaning of Swasti is – I AM Est

    Contemporary meaning is – I wish your welfare

God is a great Mystery. I AM is His name for ever and signifies consciousness, but God is much more. God I AM is Omnipresent and we all abide in Him but He does not live in anyone. He is said to be Omniscient and Omnipotent – but IS NOT Omnificent and appoints His creatures to do the cosmic management jobs

- *Besides being a Person God is also a field and there are 6 other quantum fields of I AMs with different properties. These are matter, plants, animals, humans, genius, Prophets and the last God Man in who God I AM incarnates through the Eternal Son*
- *Siva was at level 6 – Prophet*
- *Many of level 5 – who followed him also fail and this book Calls them and all others to be Saved by coming to the Right Hand of God I AM and not left to the Last Judgment and the Biblical Flood which is due anytime now*

DEDICATION

To Mary Magdalene

> She was Parvati
> Who left Siva to be born as
> Kanya Kumari the Virgin Mary
> To slay Vanasura – the disciple of Siva

Thus was she aspired by God I AM for the
End of the era of Siva and the 3nity = Babylon

> She IS the woman of Rev 12
> And Lord Kalki was born to her
> As the male child of Rev 12.5
> Who was lifted in space
> Caught up with the throne of God
> And replaced Siva

The Lord Kalki was born on earth about 18 years back
I am the only witness on human plane and testify this

This event is also foretold by God I AM in
I Kings 8.19, Isa 7.14, 66.7, Luke 1.35

PROMISED LAND

By Apostle of God – New Delhi India
Apostleofthegod.wixsite.com/flood

Self published 2016 © the Author

ISBN-13: 978-1535210171
ISBN-10: 1535210176

COVER STORY

Matt 24.30

Then the sign of the Son of Man
Will appear... and they will see the
Son of Man coming on the clouds....

> *Lord Kalki is the Son of God[1] Son of Man*
> *His sign is white horse*
> *Sometimes back this sign*
> *Appeared in clouds*
> *At several places*
> *And in several forms*

Lord Kalki was seen by the seers as a warrior on white horse with a sword in one hand. Horse is symbol of mind, the rider is the soul and the Sword of True Gnosis along with the dedicated action links the earthly mind and soul to the God I AM

> *That is the True and Integral Yoga[2]*
> *My Yoga - given by God through*
> *Narayana Krishna Kal = I AM 1 of Gita*

---

[1] God Doeth NOT beget any son – the phrase is for linguistic convenience to signify that Mystery of God was involved

[2] Also known as – the twain shall be one, that IS - the earthly mind via soul and the sword of truth becoming linked and increasingly becoming one with God I AM

FIRST RULE OF ALCHEMY

To make gold
You require some gold to start with
        In the same way
        On way of human evolution
        You first need some input Gold[1]
        From the God through the Word I AM
        Who IS in the Image of God I AM
THAT Word I AM is Eternal Son Lord Kalki
Who has now been perfected as final incarnation only

Before this True Yoga was not at all possible
Because the Eternal Son was evolving
The Yoga of past era was – of a transition age
Transition from animal origin to human form, no more

Siva lost whatever Yoga he had
Because he misused the Privilege

---

[1] In New Testament this start up gold is named as neighbor – Love your neighbor as yourself

Listen - Mankind of the Earth[1]

Siva of 3nity = Babylon[2] *was* the prodigal son of the parable of the New Testament. He left God I AM and could not make far less manage a good human creation

In broad sense the human creation was being ruled by him by the doctrine of golden calf of the time of Moses. The calf was remade from ashes after Moses left – and the race failed the test

Kalki was born to bring the new age, Prodigal Son Siva fought against Lord Kalki but towards the end he showed inclination to come back to God, Who sacrificed the by now fatted golden calf – therefore now the rich men cannot enter the paradise except they first come to the Right Hand of God through me

RECENT EVENTS

To come to God, Siva had to incarnate in a human. Siva incarnated in Modi. But incarnation in human body creates veil over eyes[3] and Siva got stuck in his own past and could not come back to God

The Bible parable did not set any time limit for return of Siva but Modi said that 2 years are enough for appraisal of his ministers, this rule all the more applied to him, and

---

[1] New Age Shema!

[2] And the Babylon is fallen today the 8 August 2016

[3] *Guru Gita – dehe chagyana sambhava*

that was Time Out for Siva team – today the 8 August 2016 onwards God requires them to follow His Word

The baton is now in the hands of PM of India Sri Narendra Damodardas Modi. This book is an open reminder to him to overcome Siva factor and come to God I AM through me – the official Divine apostle of Deut 18.18, and this reminder is according to the Word of 18.19 that requires him to hear what I say and optionally to respond positively asap

- The following rules apply to him in this instance
  - *Ask and it will be given to you. Modi has asked for India to be the Promised Land, that was not his privilege to do – by doing this he tempts God like Sonia too did, this is sin. Both them must ask for mercy of God is my suggestion*
  - *Seek and you will find. He has to seek for the door of this Way of His mercy via the* apostleofthegod.wixsite.com/flood *website*
  - *Knock and the door will be opened for you. He is required to pay $500 on the Paypal portal and would be entitled for one email with further guidelines*
- *A few years back a Swiss Psychic Spiritualist who was my mother when I was Dhruva and also when I was Sage Kapil saw several future visions about me. In one vision she saw me in Navy Blue chauffeur driven Rolls, welcomed by a man in tuxedo and a*

brunette, then a group cheered and welcomed me and I addressed them from top of a spiral staircase

- o *They could be Modi and Sonia Gandhi whose minds were controlled by Siva mafia and who tried to destroy me by blocking the Constitutionally guaranteed justice to me and who now stand on the precipice*
- o *Sonia has no future, she is weak and fell sick in UP rally, Rahul is under curse of Psalms and now Siva can not intervene to save him and he is sacrificed fatted calf of the parable of the prodigal son*
- o *Modi as the prodigal son did not return to God. Though the parable sets no time limit, the time was specified as 2 years by Modi himself*
- *I had told Sonia Gandhi to repent and return and be saved but Siva did not let her. I also told Modi and Sushma about the vision of man in tuxedo and brunette, which now appears also to be about Modi and Sonia or Modi and Sushma or Rahul and Sonia or all of these and more such pairs*
- *So I ask again by this book – I am setting no deadline because these are the days of the end, but the vision indicates that both of them must agree to meet me, in pairs. If so visit my site and pay $500 offerings and I will send one email with further instructions. This rule will apply to all kings and every other group that has been called, come in pair of the leaders and*

pay specified offering. The pair must have a woman – mother, wife, daughter, collegue who can recommend the man as worthy of divine favor and stand surety

Disclaimer

1. *Please note that in my divine capacity I do not at all recognize any nation, flag, constitution, religion, chair or person but go only by Word of God and see only those who come to His Right Hand through me and NOT those who drag heals at the left hand and must perish – as told in the Books of Matthews and John*

2. *You will accept guidance of God through me*

    - *I will guide you as God tells me but first you have to ensure that injustice done to me and my kin is compensated as guaranteed in the Constitution of India and that my status has to be above even the President of India and for that I do not require any mortal consent or writ of Constitution – God Hath Given me privilege of being the High Priest and King of the kings and we shall deal only within This Maryada[1]*

---

[1] You follow Ram and one hopes that he was named Maryada Purushottam and you ought to know significance and importance of *maryada*

According to the text of Deut 18 clauses – May the world leaders see the power of Word of God spoken through my mouth and start becoming wiser[1] – at this time I suppose I will see them all in this very way of the vision of man in tuxedo and brunette

All listed in Rev 6.15 have been called by me. The call is for two classes of the persons – slave and free. Slaves are those whose minds are captive of Siva and his falsities. The only free group is that of Noam Chomsky and Susan Sarandon and this is foreseen by Rev 8.9-17. However I expect that wwYouth Intellectuals Judges and Journalists too can easily become free if they strive

*Modi does not have divine jurisdiction to promote Yoga and has tempted the world. **Let this be very clear** that all nations must leave Modi lest they too share in his sin. This is my ruling that everyone may take his own cross and follow me – do not follow Modi or anyone else because I can save only those who COME in sincerity and not who come in shadow*

Good Luck and Best Wishes

---

[1] One way is – Fear of God is start of the wisdom

Mysteries of The Kingdom[1]

- *Matt 13.12 – For whoever has, to him more will be given, and he will have abundance; but whoever does not have, even what he has will be taken away from him*

Do you have faith in the God I AM and the Eternal Son, His first creation perfected to requirements of the threshold of the new age – if so you are entitled to receive more and also the True Yoga

But if you have faith only in temptation for Yoga of Modi or from very many false christs or those whose Yoga is vanishing by prophecy of Gita 4.2 – do read the above Word of God again and review your option IS my suggestion. Have faith in God and not in any gods or gurus of past era IS rule 1 of this age

---

[1] Matt 13.11 – Because it has been given to you to know the mysteries of the kingdom of heaven, but to them it has not been given

KINGDOM REJECTED BY JEWS

The kingdom of God was sabotaged by Siva when he worked through the Jews who rejected sons of Samuel[1]. God told Samuel to warn them of the consequences, those words are true now[2]

What is Google doing and what is doing Zukerberg? Do you think this type of digital revolution will help evolution of mankind? NO – they will sink deep and deeper in matter and virtual world and become involutionery – simply because Adam Eve and Serpent all three were cursed to go towards and in dust

Unless that curse is removed mankind can not at all go up – for that men have to remove all other sort of systems of governance and ask for the kingdom of God on earth as now in the heaven from where kingdom of Siva have been uprooted

---

[1] I Sam 8.7 – God says they have not rejected sons of Samuel but God
[2] For example I Sam 8.11, 12, 13 – such kings will take the sons of people for driving their cars as also commandoes to protect them by running before cars, will make army and police units, and other specialized groups to plow their grounds, make weapons and components of cars and planes, and daughters as perfumers, cooks, bakers and hotel receptionists

Every man go to his city[1]

This is the curse of God through Samuel. You all go back to your original city, to animals. Promised Land can never be attained by you except you revert to the Truth of God. This is the opportunity offered by this book and this is the great significance of this offer from God

Mormon Bible sets another precondition[2], first you come towards God then He comes towards you, not the vice versa. Even in case of prodigal son – Siva and Modi started coming then He acted...but they did not come and were rejected by God. Third condition is faith; you come in faith because no one has seen God and no one can prove Him. Alas Modi tempts God by his flowery talks – either of future or of past. He is the piper taking the rows of the lemmings to the sea of cold death

---

[1] *This is the last sentence of I Samuel 8. And this is a Big Curse – this means you go back to the animal City from which you came and lick dust – this curse denies you the Promised Land*
[2] This is written elsewhere in this book

COMMANDMENTS

The 10 commandments of Moses apply in the new age

> The 2 commandments of the New Testament are Love God and Love thy neighbor. Here the neighbor is Gold of Word given in the process of Yoga. For Mary Magdalene this Gold was named Lamb, of - Mary had a little Lamb nursery rhyme

> These two commandments are found in
> **Vorpal Sword = Love God I AM + Word**

These 2 commandments do not apply to average persons. The first commandment applies only after he enters a required stage of Yoga with God. The second to who the Gold particle of Word is given in the process of Mysteries of Yoga

To love the Word and God are not academic far less lip service but subject matter of plane of Divine Reality – that is exactly why You Start with Faith

FAITH

Bible says that to enter the kingdom of God one must have faith of a child

But faith is also a double edged sword. Faith of a child can also be exploited by devils. Ganesa was created by Parvati. She told her to guard the door. Siva came and Ganesa stopped him. Ganasa had faith of a child. Siva did not like to be challenged so he beheaded the child – and later fitted head of a demon named Gajasura. Parvati was fooled and shown down in this tricky political game by Siva

Now Modi plays tune of the legendary piper. The world follows him NOT as child but as hypnotized adults. They must overcome the black magic and be saved by faith in God. I am giving true logic in this book – any child can understand that, but if one comes with pride and prejudice he cannot see the plain truth

❖ SIVA FAILS

Failure of Siva was set when Adam was cursed to go back to dust. As prodigal son Siva was treated with mercy by God but he threw away the offer courtesy the ignorance and false pride in 56" air bag of PM Modi. The option of repent and return is even now available to Modi – I add this paragraph on 16 August 2016 and plan to finish this book very soon

Yoga is for humans. Siva did not come on human plane but spied upon Yoga of God men, even stole[1] from souls and gave to his men, specially to his snake Patanjali and sent him by a trick on earth when Aniruddha the Grandson of Krishna and Usha the daughter created by Parvati decided to come on earth. Siva feared that they may attain Ardhnariswara Yoga which he had failed to do in Amarnath. The snake Patanjali would have stopped Aniruddha and Usha by creating hurdles if they tried to attain that status. Siva did not try to think that he failed because he let Sati die and God did not chose him for that status. Let this be inscribed on Granite plates of brain – True Yoga depends on God and not on technicalities which are secondary

---

[1] Phoenix of Osiris the Satyavana was stolen by Yam, even as he fooled Savitri and delivered her to 100 sons of Dhritrashtra, like as Yudhister he gambled away Draupadi to them. This Phoenix was later collected by Gandhi from Africa and given to Aurobindo Ghose = Judas Iscariot also Ameeta Mehra = Babylons

THE CHANGE – TAROT CARD X

This tarot card is Prophecy of the Change that is on the cards these days – the Sphinx is rotating the wheel of fortune named Rota

The snakes are going down the precipice as the wheel rotates. Those who rule are evolved as having human head but being of animal origin they are having rest of the body of crocodile. They are next to fall – except they come to the right hand of the God, so that God redeems them from the animal origin mind set which Siva could not or at least did not do. If this class does not LISTEN to God the next race that would replace them is shown as dog. Not yet having human head of high wisdom but they may be loyal to God and when they rise or even now when the call is being given by this book to the wwYouth

These prophecies were inscribed on the entrance door of the way of Initiations in the Great Pyramid and destined to reveal their Truths in due time

I was Pharaoh Atothis in the book Initiation by Elizabeth Haich. My elder brother was Ptah Ho Tap the High Priest secretly aligned with Siva and a cheat. There were many students but several of them became my enemy Yogs in this age – the leading one was Raman Maharishi of Arunachal. One student was born to Elizabeth Haich, his name is Gideon Haich and a US University is doing research on his past birth in Africa. I wish he joins the mainstream of new age Yoga – inform him if you know him

SPHINX

Sphinx is made of 4 creatures. One is the bull who was symbol of Dharma as preached by Siva, another is lion the Nrusimha incarnation of Eternal Son Narayana, third is a winged bird named Phoenix who was born from the ashes of Osiris – this signifies the drive of human soul towards God and the fourth is human with Wisdom in Cerebellum

These four faces are also part of the visions by Ezekiel and John of the Rev Book

Sphinx Riddle

Sphinx guarded the way to Pyramid of Initiation and would ask a question – who walks on 4 in morning, 2 in noon and 3 in the evening. The answer given by Oedipus was a human

New Age Meaning

In this era the lion is Biblical Sher G which role I played for some time, the bird is the male child of Rev 12, the snake represents Patanjali and many others who came to oppose Lord Kalki, human head crocodile - as one example, is CJI Thakur who shed crocodile tears before Modi and public in a conference that is indicated by Isa 1.18 to have been with me and not with Modi. Most rulers of this era are such human headed crocodiles. Unless liberated from animal past they fall down the precipice therefore must they Come to me. The human head signifies the perfection of Narayana as Lord Kalki. The human age starts now

One who walks on 4 is animal or snake, who walks on 2 is primate, one who walks on 3 is the person with Staff of Life where staff stand for spinal cord as snake risen up towards God. In this way the Sphinx puzzle indicated the broad sequence of human evolution. In spiritual yoga of era of Siva the God was not involved because Siva had left home as prodigal son, and though the Kundalini or coiled snake did rise, it remained same snake. The oblique purpose of Siva in large scale

Kundalini awakening was to make an army of Yogis to fight against Lord Kalki. The very process of Yoga was based on entry of a mantra or ray from the eyes of guru in the land of the mind of the disciple as a Raktbeej to make one more person in likeness of demon Raktbeej – this is clearly shown in what Smith did in movie series Matrix. Much of this Yoga was destroyed in a long war and towards the end the Anna Hazare episode of caps was an illustration of a desperate Raktbeej phenomena

GOAL OF HUMANS

In era of Siva the goals of humans were Bhukti and Mukti – that is, enjoying the world and liberation. Both were false, few persons at top enjoy true happiness was told by Leo Tolstoy in his story of the shirt of a happy man, and their fall is in eternal hell and there is no liberation as hoped by Siva. Bible rightly says – wages of sin are death - from human zone and fall from the wheel of fortune

In this era the goal set by Siva was – *be ye perfect even as I am perfect.* This call came from God through Siva circuits. This call is valid and this perfection is equal to the Eternal Son who is the first manifestation of God. But looking from point of view of Siva or Raktbeej this very call has entirely different meaning

Here you must also add the law of similarities, this call appears to be open for all, and yet only one had to succeed after that the race is over – like several million sperms race to penetrate one ovum and once that is done by the first sperm the race is over

Call from God was meant for this event. But Siva took it differently and tried to make his RELICATES as in movie Matrix or the legend of Raktbeej. Nirmala devi clearly said that she would make every one disciple as Ganesa. The same objective was meant by the name Ganespuri. But this was not the meaning of God. In his greedy blindness Siva could not add 2 + 2 as 4

This gives a very important teaching. Sincerity unto God is the highest human quality. No one knows God so one must start by being sincere to his own self

Here is a small counsel to Modi. In politics of India we often hear the great Indian leaders that one should not do politics in cases requiring human ethics – the question is that if politics is so bad, why at all do politics? Modi will not read this, at least he will not answer this question – but others must ponder upon this. New age will be suited to those who have hearts of a child – they may ask what is sincerity...

Modi errs. The die is cast. Only one man had to be perfect as Eternal Son. Only one man had to get the Holy City and the living Temple of God. There is now no vacancy at top. God offers full savings because men at top were fooled by Siva and per se they did not do specifically personal sin, at least not 100%. They must come out of past mind set and be saved

> *God came as robber and told Modi – your life or your wallet. Modi said – take my life because money is for my retirement...Modi do not err to retire in animal zone forever!*

In this back ground must you interpret the Biblical phrases: Come out of Babylon and Overcome. My Job as apostle of the Deut 18 is to tell the truth. It is up to you to overcome past mind set and come out of the fallen Babylon, and to come to the Right Hand of God if you do not want to fall for ever from human zone – and *this IS the purpose of this book*

I. PROLOGUE

3NITY = BABYLON IS FALLEN

Old Testament says that I AM is the name of God for ever and His memorial for all generations – question is, would that end the global conflicts in the name of God?

> *The question is wrong because the basic problem lies elsewhere. The old cosmic rulers of the 3nity = Babylon were deliberately mismanaging the mankind to prolong their own tenure and contrary to the Will of God I AM*

This was done by many magical tricks – one very evident trick being invisible control of human minds. One clear example of this was seen when the US Congress gave 7 standing ovations to one speech of that Modi whom they had banned for entry in USA. This was a clear case of mass hypnotism

Here is one more clue. Has any Indian ever wondered why Siva used to give very strange boons? Most devotes asked for immortality, the rulers of 3nity would refuse that, the truth was that neither they themselves were immortals nor at all able to grant that boon. Always a roundabout boon was sought to fulfill the same objective. So long that worked – the giver of boon also enjoyed longevity. This is how Siva wanted to be immortal through Ravana and next through Vanasura

So long you believe that Achhey Din will come, the tenure of Modi is guaranteed. Modi knew that Achhey Din will come in the new age and used this to grind his axe and he is followed by those who have their own knives to grind. This is the basis of Hindi proverb – Ram name in mouth and sharp blade behind the back. In contrast the person who revealed the Rev Book had two edged sword in mouth and God hidden behind his back. The two edged sword is Sword of True Yoga

HEAD TRANSPLANT

Siva the chief of the 3nity knew the science of fitting a new head, once he fitted a goat head to Daksh, next an elephant head to Ganesa

Modi says that these were head transplant operations in the ancient India – but no, they were examples of changing the Mind Set of any person by abuse of power of Siva

See how he changed HEADS of US Congress. Once they won't even give Visa to Modi, next they would give 7 standing ovations to his speech which was neither spoken well nor drafted. Main question IS what that speech gave or has potential to give to anyone? If nothing why so much applause? First Obama worships black magic god Ganesa in White House next Siva comes to deliver speech and everyone thinks that Achhey Din are coming. How come US Presidents take oath on Bible and no one sees that Deut 13 damns Obama and Washington. How can Modi bring Good Days without first revoking those Words of God? OR why at all take oath on Bible and not on the Book of speeches by Modi?

> *There is more to this –*
> *Siva later restored the human head of Daksh*
>
> *That is to say – the head remained of goat but given a deceptive human mask*
>
> ***This WAS most sinister, as we now deal with such deceptive humans at the top of this world. Man Mohan Singh was a typical goat with human head, also the blue turban antichrist of Nostradamus***

Era of Ardhnariswara, Adisakti Siva and 3nity Durga and even of Krishna is now past and I skip the details to converge to the main objective of this book, to offer the True Yoga and thus save the mankind

The new age religion must prevail at the top of mankind – and the God is within. One has to start from his or her own I AM, cleanse and evolve that. The 7th level I AM links by choice of God I AM and that can never be attained by any technicalities – this is the simplest and biggest Mystery of True Yoga

II. God I AM

On 29 July 2016 the demon Bhandasura[1] was slain and the Red Queen and the Knights were expelled to the Outer Worlds

- Earlier the Rev 17 Mother of the Harlots and the beast of 7 heads and 10 horns[2] were slain

On 31 July Siva and his select team who were in hiding contacted me to shake me off the side of God I AM, unsuccessful they surrendered to His Will...later I learnt that Siva and his team had made many replicates as in Matrix movie

All things are being made new. God has marked 144000 persons to be saved and a large number which I am not allowed to reveal to be saved as base of the pyramid model of the new age populace...later I learnt that God decided to save only 144000 because Siva was not cooperating and was pressing only his agenda

- God I AM has asked me to write this book

To call everyone else to be saved by coming to His Right Hand and enter the new age – Welcome!

---

[1] The demon slain by Alice was Jabberwocky = Ash 2 Bhandasura
[2] This beast was related to - Rev 17 - dying Yoga of past era. The 7 heads were presiding deities of the 7 Yoga centers and 10 horns were 5 each of the sense and action organs

GOD I AM HAS SENT ME TO YOU

God I AM told Moses[1] - Say to the children of Israel that God I AM has sent me to you, and this is My name for ever and this is My memorial for all generations

Moses told the king of Egypt - ...please, let us go [2]three days journey into the wilderness[3], that we may sacrifice to the Lord our God

Do you know what is wilderness = a-z I U Babylon?

When a child is born he or she is like a tree, getting food through placenta. When a new soul enters the child and he becomes alive and self conscious of existence, first he thinks that he and his mother are SAME. After a couple of days his eyes get focused; he starts seeing the reality and learns that he is he, mother is mother, father is father, and cradle is cradle. In ancient Indian texts this is called Ahamta Idamta   I ness and That ness, or Prakasa Vimarsa – light and understanding. This is I U field but not yet linked to search for Him

Human mind has a homing instinct to seek Him as the Origin – the God I AM. The way of mind to That I AM is

---

[1] Exodus 3.15

[2] 3.18

[3] Wilderness = a-z I U Babylon

through the sub conscious mind which is also named as transmigrating soul. That is more confusing WILDERNESS and if a person is not guided by a True Guru – he gets entangled in some past birth and falls back in time – that is how Siva failed to lead Parvati to immortality. He could not explain why Sati died and Parvati saw through his complicity and decided to leave him to be born as Kanya Kumari the Virgin Mary. Siva colud not have led the mankind ahead far less to Eternal Life, because he was smearing them as ash on his body, making them alive and recycling again and again[1]

---

[1] Also using them as biological batteries for use by his spirits as suggested in the movie Matrix. Siva used 51 parts of ashes of his wife Sati as spiritual battery. Later I learnt that he used heads of Ravana and arms of Vanasura in similar way

PRODIGAL SON

All concepts of God were given by the ruler Siva and his
government

> Siva was prodigal son
> He asked for his portion of goods[1]
> Took, journeyed far, wasted his possessions
> With prodigal living

Having thus gone away from God I AM
For his own self-rule Swaraj
He never told truths about God I AM to mankind
He always projected his own self
As the ultimate god for ever

> He knew that his era will end
> He tricked any and everyone
> Who may rise to replace him and
> Thus he hoped to perpetually prolong his era –
> But he failed

---

[1] Luke 15.12

TRUTH WILL LIBERATE YOU[1]

In era of Siva
You worship him or his deputies
And were given good fortune
As a return for selling your soul to him

- TRUTH will set you free
- *This IS the Great Promise of the Book of John*

Q: IS MY I AM NOT THAT I AM?

You are right. There are 7 observable fields of I AMs – ultra matter, matter, plant, animal, human, genius, gurus. They all have typical quantum field characteristics of the respective I AMs. I am working of deeper details and some of these fields do have 7 more subfields. Of the 7, the last field of gurus became a Rebel and is named prodigal son in the New Testament. This is where Siva eventually failed. By cascade effect the genius humans too went astray. The Siva effect was applied to their minds as - Human on Face but Goat within the human masks. No one from this field knew that One from this field may challenge and defeat the next level IF He bypasses that and links with the next field of God. But the dialogues of Matrix movie indicate that they knew this

---

[1] John 8.32

**God I AM is different than all these I AM fields –
where the quantum I AMs are embedded with field of
primordial matter**

**God I AM is sometimes named Pure Consciousness
but is more than mere Consciousness and IS a Great
Mystery...forever by mathematical inferences**

**Other I AM fields are found as Mixed – consciousness
with matter**

*Here a little explanation will be very helpful to the true
seekers of truth. All mixed levels of I AM have
consciousness SHIFTED from soul to brain and is socially
structured, that is – gets knowledge as taught to the
brain by the teachers of that social group. This is
conditioned knowledge and never the true knowledge.
Siva took advantage of this, gave different concepts of
God to different social groups and that made them fight
among themselves. Siva then blames mankind before God
to prove his point that God should not have created
humans at all*

*The obvious solution was to bypass Siva of level 6 and
link with God I AM of the 7th sky – but that cannot be
done by anyone, only God Chooseth whom He Chooseth.
In India Sonia Modi Anna Kejariwal do not like this and
trying to prove that God erred and must Choose one or
all of them instead of me. Sonia tried to tempt God by*

showing Him that she brought RTI, Anna came with theme of Lokpal as if that is any solution of anything and Modi comes with a bagful of shiny toys. No one wants to hear what God says through me.You all follow me or get out of the Way. Deut 6.15 says that you may even be destroyed from the face of earth, that is, you may go to animal zone on new planets. Dan 2.235 also indicates this – the wind will take away the crushed statue, it may take the particles to new planets where evolution may be starting from scratch

JOHN 3.16

If you believe that God I AM loves the mankind, and so much loves the mankind that He gave His First Creation in His Own Image to you and *IF* you believe in Him....

> Then by True Yoga
> The inner patterns of your organs and cells
> Will become in That Image

That is why you would get Eternal Life!

Eternal Life is a reality and not mere academic jugglery of words. However in past era many essays have been written on this subject. I cannot do that because God leads me step by step and I am unable to explain what Eternal Life is. I know much of what God Hath given me – but that is between Him and me, not that I understand all of that as yet

Even so – I will give you some idea of Eternal Life. You do not know what you were in past life. In next life you would not know what you are in this life. Thus – you only live ONCE. The Soul lives several lives, abiding first in someone then in you then in someone else. You are being used for education of Soul. Eternal Life starts with engagement with Soul followed by marriage

MARK THESE TRUTHS

- If you believe in the First Manifestation of God I AM as the Eternal Son Lord Kalki and come to the Right Hand of God – this is start of your journey on the way of Eternal Life

- *In contrast, when Siva married Sati[1] he set forth a marriage vow that IF she ever disbelieved him, he would leave her. This was a ploy to always keep her under him. BY complicity he could never have attained the Ardhnariswara status which he was after*

---

[1] Who represented Prakriti the Creation

THE CALL !

The kingdom of God in heaven has now been established by Lord Kalki who was born on earth and lifted up and who replaced the Siva. That kingdom is now increasingly coming in me so that the kingdom may come on earth and reach YOU

And that God I AM sends me to You through this book. Earlier Moses has asked the king – Let my people go. Today I ask you – Come out of Babylon lest you share with her sins

COME !

Come! And let him who hears say Come!
To his kin and friends
And also know and
Believe
Even a size of mustard seed in Him
And that

BLESSED IS HE WHO COMES IN THE NAME OF GOD I AM. COME! AND BE BLESSED!

❖ ALICE IN WONDERLAND

Alice in the Wonderland is a NWO prophecy. The sword is that of Yoga and was actually given by me to someone and robbed from her by the Siva mafia. Alice was growing like a spirit child similar to that of Rev 12 but was attacked by Siva mafia and an abortion was forced. The tree and the hole at the base of tree indicate this unfortunate event. God sent a Hatter in the earth who saved, trained Alice who retrieved the sword of White Queen the Kanya Kumari and slew the demon Jabberwocky = Ash 2 Bhandasura – the love god made from ashes after Kamdev the original love god was burnt by Siva. This is another major problem of human evolution – their love is like someone burnt and remade. Secondly that god is now dead. Modi cannot connect to any of you to new age god of love – therefore leave him and come to the Right Hand of God

I. VORPAL SWORD OF THE ALICE

VORPAL S = LOVE GOD I AM

This is same as the first Commandment[1] of the New Testament

VORPAL SWORD = LOVE GOD I AM + WORD[1]

---

[1] Matt 22.37: You shall love the Lord your God with all your heart, with all your soul, and with all your mind

*This is same the first and second Commandments[2]
of the New Testament*

VORPAL = A TWO EDGED...(REV 1.16)

*The phrase implies that this Sword does both functions –
to save some and to finish off some. The Alice book
clearly says that the Sword knows what to do...In Indian
books too Yoga is told to be a sharp edge of a knife, now
we know that this refers to The Vorpal Blade*

- *AND THIS BOOK IS ABOUT YOUR SAVINGS!*

A NU VORPAL BLADE = SWORD OF KALKI

After this revelation God I AM asked me to write this
book of True Yoga

ALICE IN THE WONDERLAND

By wiki - Alice's Adventures in Wonderland, commonly
shortened to Alice in Wonderland, is an 1865 **novel**
written by English mathematician Charles Lutwidge
Dodgson under the pseudonym **Lewis Carroll**. It tells
of a girl named Alice falling through a rabbit hole into a
fantasy world populated by peculiar, anthropomorphic

---

[1]In these two the definitions of the Book of John apply. In the
beginning was the Word and the Word was with the God and the Word
was God. No one has seen God

[2] Matt 22.39: You shall love your neighbor as yourself. Neighbor is a
starting Gold Particle from Word given to you as the Divine Mystery of
Yoga, there is no True Yoga without THIS

creatures. The tale plays with logic, giving the story lasting popularity with adults as well as with children. It is considered to be one of the best examples of the literary nonsense genre[1].

> *Lewis Carroll said that the names in this book were made by mixing English words. When asked about the Vorpal Sword and Vorpal Blade, he said – Not Yet. In hind sight we see that God I AM spoke through him*
>
> *One Mr Taylor suggested that Vorpal is made by alternate letters of Verbal and Gospel. This confirms the Mystery of God I AM and the verbal algebraic equations in this book, Bible and other books; and that the Alice book is a Gospel*

ALICE – A NWO PROPHECY

I testify that the events of this book actually happened on spirit plane in the recent times and I was made aware of the major events

Alice actually slew one Jabberwocky quite sometimes back in USA as a practice run, and very recently[2] the ROOT of Jabberwocky was slain under the earth...

---

[1] We now find that this book is a NWO prophecy

[2] On 29 July 2016

Jabberwocky was god of love and had a ROOT, a field and every male human had his own Jabberwocky

JABBERWOCKY = ASH 2 BHANDASURA

Present creation by Siva was based on several Ash Factors, here I show only one – the Bhandasura

THE DEMON BHANDASURA

After death of Sati; Siva made another wife for himself from ashes of Sati at Varanasi. She is named Parvati because she was born actually as daughter of a king of region of Himalaya Parvata

> *If Parvati too rises and poses a threat to supremacy of Siva, and requires being finished, Siva would have made others from other parts of the ashes of Sati which were guarded by his commanders. In real life, Parvati left Siva and was born as Kanya Kumari the Virgin Mary. But I saw Siva with a wife in Ganespuri – they were about 60' high, stood on lotus flowers and I saw them with open eyes. Later I learnt that he had made the new wife from ashes named Vajreswari. Thus Siva played a big trick by dividing ashes of Sati in very many parts – this trick was a conspiracy of 3nity, either 3nity was aware of this or Siva induced thoughts in the minds of other 2 to accomplish his plans*

Matrix movie says that he used mankind as biological batteries for his machines, but here we note that he turned to ashes his divine wife and used the ashes as spiritual batteries for men who were as sycophants to him

The error of this prodigal son was that he did not turn back to God I AM but tried to show Him down and to prove that he too can become like and as powerful as Him. He forgot that when he had spoken a lie this God I AM had snipped his upward looking head[1]...

> *Technically speaking, after this, Siva would never had become loyal to God I AM. From another stand point – now he can descend to do creation, and being unguided and free the creation will show Good and Bad and once that is known, the next phase of creation can be the Tree of Life where the new rulers will have advantage of hindsight of what to do which is Good and what not which is proven Bad[2]*

[1] Indian legends name Brahma as the one whose 5th head was sniped off, but names at different ages were deliberately mixed, I have to skip huge details in this book which is but a First Primer for mankind who know very little Truth

[2] THEREFORE Swaraj or self rule cannot again be given. Why to reinvent very many prodigal sons as if one has not done enough harm to the fabric of creation

Siva was caught between two opposite ways. He almost became schizophrenic. On one hand he wanted to get rid of his life which started with sins – therefore he preached Liberation from Life[1]. On the other hand he wanted to live and to sit at the top Chair for ever. The psychology of the men at top of human society is similar

Siva did not want to marry Parvati though she was created for this destiny and had done long Tapas to get him. Long Tapas ensured that the person will be loyal to Siva forever, and yet Siva was hesitant. But Brahma wanted that the circus must go one – and sent Kamdeva the god of love to prompt him out of his meditations. Siva did come out of his trance but burnt the the love god to ashes

Brahma remade love god from ashes
That god WAS named Bhandasura
The very human love is remade from ashes and this is clearly seen in the human behavior. This situation had to be changed

The change required end of Bhandasura and connection of new age god of love with the human

---

[1] Moksha

circuits. We now enter in era of True Human Love – to say more will be premature, first you have to be redeemed from the ditch in which the mankind has fallen and the kings of the nations must therefore hear and respond

The problem gets complex because spirit mafia of Siva enters humans to create chaos everywhere to prevent the Change. They cause cruelty on mankind on several levels – one to one plane or organized violence by State. They also cause creation of masked human VIPs talking Good but who are hell-burnt[1] against God I AM within their hearts. In this situation anyone will say that those who are violent are certified Bad and those who talk Good are Real Good. But both these are mischiefs of Siva team. If the violent ones are removed by Lord Kalki his other team remains with laurels - this is the depth of Satan

Humans do not see the root cause of events and classify persons as good and bad. But my research shows that mafia of Siva indulged in cruel gang rapes in India because of their frustrations against the warring Kalki...I skip details because there is so much objective work to be done and mankind is in mode of only talks and not

---

[1] Yudhister has interceded for all who died in Mahabharata war and went to hell, those hell-burnt sinners are creating big problems to mankind

capable yet to do anything to save the situation – I hope some may rise after reading this book. To RISE seems mandatory if one wants to inherit the earth – to lay down and watch or not even that does not seem to be the preffered way of doing that

God has tested and failed Siva and his management by my case of Guaranteed Rights. Indira, Congress and Judges cheated me, then in times of V P Singh, BJP stopped him from starting impeachment proceeding against the Judges, then Atal Behari cheated me, now Modi is cheating

Modi wants me to fight against Sonia and also against the Judges so that by default he becomes the hero and intervenes to help me. But not I – he needs my help. I have become what God wanted me to become and am increasingly becoming more of what He wants to make me but Modi is falling every moment because he is incompetent and hollow

In my case the main culprit is HAL who also played fraud with Judiciary – by their consent. Media too is part of the conspiracy. This was all handiwork of Siva which started by Emergency to block my Truth and implicated all top of India. Modi proves himself impotent so far the issue of Justice is concerned. Justice is the main theme of this age and Modi falls on face in this context

II. Mysteries of God

I testify that I witnessed the following events in last 10 days of July 2016

A. Alice slew a personal Jabberwocky in USA in 2014 and now the Mystery of God slew the root demon under the earth who made a field linking all personal Jabberwocky = Ash 2 Bhandasura – the love god made from the ashes of Kamdev burnt by Siva

B. Soon after the Red Queen and the Knight were expelled

C. A few days later Siva defended himself in final Judgment but could not hold grounds. Like Smith in the movie Matrix, he has made many images of himself and they were wiped off. Later I found that there were more and more in hiding and they were wiped off in stride

4 August 2016

III. Secrets of Yoga of Alice

Alice learnt six secrets during her adventures in the Wonderland – they are all Secrets of True Yoga. I will explain here only one or some of these, others you will learn when you come to the Right Hand of God

1. First learning was that she could grow by eating a cake

2. Second that she could become smaller by drinking a potion

In Indian Books both these are written as – small as a point and as large as sky. These are also known as two special privileges[1] of Eternal Son Narayana that can be attained by True Yoga

3. Animals can speak

The evolution is from unconscious matter which is animated by God I AM and the Eternal Son. First animation is fivefold named Panch Bhutas...which are the first 5 centers of True Yoga. The old era presiding deities of Yoga have now been changed[2] therefore old Yoga will perish. All presiding deities of old era have been replaced or are being replaced – this is why one must come to the Right Hand of God

---

[1] Siddhi

[2] In fact the Presiding Deities of all 7 centers have been changed. In Rev 17 they are the 7 heads of the Beast. The 10 horns are Deities of 5 cognitive and 5 action senses – they too are changed. The Mother of harlot presiding them is gone. All 330 million Deities have been changed by God I AM...that is why one retains human life only on the Right Hand of God I AM

4. Cat can disappear

This mystery cannot be revealed here. As one aspect Bible says that people may be in festive mood when the Flood comes. This means that the money intoxicated men would not be able to see the cat – Modi may see the wrath of nature in India and else where but closes his eyes to the BILLI by doing all sorts of diversionary festivals unaware that the Cat may appear anywhere when she so wants

5. There is a Wonderland

Mind sees only the wilderness = a-z I U Babylon that too only of the conscious mind, but knows not the origin and reality of thought – which is Wonderland. In Yoga one ought to have control on thoughts but more cannot be revealed here. The highest attainment of Yoga is full alignment with the Words of God I AM which is most wonderful of all wonderlands

Later I found that names Alice and Wonderland are also spoken in the Matrix movie and the way God organizes all this is a Great Wonder for me

6. I can slay Jabberwocky[1]

- *This is the Most Important secret of new age Yoga. The mankind has love controlled by a demon named Jabberwocky who was made from ashes of the love god Kamdeva who was burnt to ashes by Siva...see around and find examples of burnt and remade love, for example*

1. *Modi is proud of India and Hindu religion but would like to force temple of Ram when live temple of Lord Kalki is on the earth. His love of Ram is burnt love – love of the ashes of Ram*

2. *When editing this book there was an exposure in India about duplicity about Sacred Cows – even Modi said that 70 -80% Cow Protectors were actually gangsters – burnt love to Holy Cow!*

     - *Later I found that a TV Channel did a sting operation on this issue and Modi spoke a day or two before that. Obviously the sting operation was not done in a aday and in reality the fact was leaked out from the TV team so Modi did a premature damage control and thus was caught, let us say – coat down. Coat woven of rays of Sun Moon Stars. Modi has moles in media! But he loves only his own right of freedom of expression*

---

[1] *Jabberwocky = Ash 2 Bhandasura*

3. *Modi himself is up to nose in filth of Injustice in my case and drowns my voice by his nonstop loudspeakers – what if God may hear my cries? Why must Modi not ask HAL to confess before Judges? Simply because Judges too were part of the conspiracy?*

4. *American Presidents take oath on the Bible on which George Washington took oath. But Obama does not want to know that Deut 13 damns him for worship of Ganesa in White House and US Congress does not want to know that they too are damned for supporting Obama and 7 standing ovations to Modi. All claims of love of Truth Justice Equality are that of burnt love*

WONDERLAND

I really do not know what is meant by Wonderland of Alice. Very possibly the visions given by God I AM to Neo and also to Alice fall in this definition. I know this Wonderland, I know a very few persons who know something about wonderland – but I doubt if others can understand it

I know that Jews fought with their lives to come in the Promised Land. Some settled in Israel, others thought that US may be the Promised Land

But this is not about LAND, this is about Temple in my human body and the followers of this High Priest who I AM. The Promised Land is not across the stones of Jordan River, but the river and stones are secrets of True Yoga

This places some burden on the Jews to come to me first. The Jews were persecuted by Hitler who was Gabriel and took revenge of crucifixion of his father Michael. But those souls came to India. RSS Hindus imitate a salutation to Hitler. Fearing Hitler they wish to be like him, rather than be like the persecuted ones. They are burnt at core, having a fear psychosis within. Only I can save them – on terms of God and never on terms of Atal Behari or Modi. But these Hindus were captive of Siva and being used by him. The Jews of Israel, though in the same lineage, are different souls. Of them Larry and Page shows great courage and foresight. A decision was made to copy all the books of Stanford Library for Goggle data bank. The books were cut at the binding, photocopied and rebound again. In contrast Modi only talks

DNA and Souls are different. Souls may evolve irrespective of DNA was proven by Antipas = Bhim Rao Am od Rev Book but that does not mean that Mayawati can equal him because she belongs to the backwards not that Modi can equal him. A man excels only by God

I sense new age potential in Jews and expect them to come forward. I especially liked the childlike simplicity of David Vise in his Google book. I note that the movie Matrix is about those humans of HIDDEN Zion who are searching for the new Moses of Deut 18. Matrix movies have a very high level of Truths of Wonderland and now that machines of devil have largely been exterminated I expect them to contact me – the Morpheus, the Oracle, the Senator, the Chairman of the Senate, Niobe...the movie is real so must be the characters – Come

I have noted work inspired by God in several Journalists and media teams. Devadutta Pattanaik showed many truths of Siva from his own point of view but God led me to find hidden clues. Epic channel researched legendary truths in India and in case of Krishna and Ruksha-mani gave me the exact information that was meant for me. The latest was the episode on Radha after Krishna left me. In that I could pinpoint if Krishna ditched Radha or Siva fooled Krishna to leave Radha – this was very vital for the final Judgment because I had to Judge my Guru Krishna

In this context I found that ashes of Krishna were misused by Siva to make several lower plane versions of Krishna like Vitthal and all such moves had a sinister purpose of the prodigal son Siva

In hind sight it is easy to see that Siva smeared ashes of cremation ground to remake those men because being prodigal he had spent his portion and was recycling the matter

Let me add that enemy had made me desolate. BY some miracle someone taught me how to work on internet in DU facilities for the staff. Then Google added features which were required by me…events were and ARE under fullest control of invisible God I AM who leads me, without That I was dead thousands of times by now…Siva did not take cue even from that

And yet friendly human minds are not coming forward because they do not try to overcome their habits of worship of Siva. Obviously God does not want to use magic to coerce them BUT waits to see who can overcome the outer sensory and socially structured Gnosis by impulses from inner Wonderland – I show LIGHT, this is the right way compared to burnt love of Hindus with RSS and BJP

❖ MYSTERIES OF YOGA

Mysteries of - Yoga by the God of Humans must the better be directed to evolution of men and society and not merely to bending the spoons. It is pity that in India miracle of any sort is considered divine – Siva and Ganesa both were black magicians and men revered them because of their miracles, power, fear and long indoctrination by devotional prayers...no one ever asked why Siva let nations fight in the name of God, and why men prefer statue gods who cannot point a finger towards their sins, thus a corrupt Indian politician or a businessman starts with divine yagya and fearlessly indulges in corruptions - thus protected by gods, few Hindus know that yagya is never in conventional fire but...in the Fire of God!

SHERLOCK HOLMES

When Sir Arthur Conan Doyle expired and went to paradise he was told of the big fraud played on the mankind by Siva. He said he want to contact someone on earth and tell this. God allowed him to contact his wife. She was in a photo session facing the old type box camera and face of Sir Doyle came in the inset at top of the photograph. He could not contact more than this. When I say that photograph it became clear to me that more than 1000 miracle photographs of Sahaj Yog Ma

Niramala Devi were nothing more than spiritual tricks – that changed the perception, courtesy Sherlock Holmes. Since then I have known hundreds of calculated miracles of God and they are all aimed to bring kingdom of God on earth for the Good Day for humans – claims of Modi to do the same are doubtful because he does not have the spiritual means even if the intentions of which there is no certainty

> *There have been TV reports of apparitions on UTube and investigations by Paranormal Society of India – I suggest they study the deceptive miracle photos of Nirmala Dev*

I. THE RAINBOW OF GENESIS

By Gita, with end of the creator his creation also ends up. But God I AM promised in Genesis that He would never let whole creation be destroyed from earth. Therefore He has marked 144000 persons to be carried forward in the new age, another large number which I am not revealing also enter the new age in next circles[1]. Others must link up with Me and that link is the RAINBOW which God I AM will see from heaven[2]

---

[1] Later this was canceled by God. The full story is that I had asked Him to save many more and He agreed but later He showed me the reality of the top men and suggested that needless mercy may not be extended

[2] Again, this is symbolic text. In reality That God is everywhere and I am sure He knows what is going on

COME OUT OF BABYLON

This book calls for all others to come out of Babylon and be saved. Because time is less the call is addressed primarily to the kings of the nations who must come with the honor and glory of their nations, however private groups may also come irrespective of the decision of the kings

II. MYSTERIES OF YOGA

1. BHAGWADA GITA

Bhagwada Gita means Song of God. Shortened name is Gita. There are more than 1400 editions of Gita in 34 languages

Being the Word of God, Gita is very powerful. Why then the mankind suffers – is the most logical and important question of this time!

2. YOGA OF GITA

Gita has 18 chapters, each named after one type of Yoga. They are actually steps of Yoga and the complete teaching was for Arjuna – not for everyone else

3. TYPES OF YOGA

My Guru Lord Krishna said in Uddhava Gita that Yoga is of 3 types and there is NO OTHER Yoga

1. Yoga of Gnosis
2. Yoga of Action and
3. Yoga of Devotion

These are not independent but combine in many forms and are stations of the Way of Divine Human Life

One can derive Integral Yoga[1] by combining all three as - the Gnosis bestowed by God I AM is put to Actions for doing His Will. This is only for those who get this Job from Him

4. ORIGINAL YOGA

Gita 4.1

This imperishable Yoga was told by God I AM to Vivaswana who to Manu Vaivasvata[2] who to Ikshvaku[3]

5. ORIGINAL YOGA LOST

Gita 4.2

Thus transmitted in succession from father to son this Yoga was known to the Royal Sages, but was lost to the world after long lapse of time[1]

---

[1] By default the Integral Yoga of Aurobindo Ghose = Judas Iscariot was a downright fraud of the past era. The Supra Conscious of Aurobindo was actually the Phoenix stolen by Yama from soul of Osiris the Satyavana, Gandhi went to Africa to bring that to India and give to GOI protected Auroville – for conspiratorial *tryst* of Satan mafia with destiny

[2] He was Yama, Yudhister and Lord of Dharma failed by repeated misconducts. When as Judas he sold Jesus, Narayana said that it was better that he was not born at all

[3] Ancestor of Ram clan

6. ORIGINAL YOGA GIVEN TO ME

Gita 4.3

The same ancient Yoga

And this Yoga is Supreme Secret

Today I tell you

Because you are my devote and friend[2]

7. SUPREME SECRET YOGA

The above Yoga given by God I AM through my Guru Krishna to me is Supreme Secret of God. BY another meaning this Yoga is Good and Secret. God said that Yoga is Good so that I trust Him and do not decline. Today when I am preparing to impart the teachings to who qualify – I tell you, This Yoga is Very Good[3]

---

[1] This Yoga is NOW being finished off by Mystery of the God I AM Who IS the Master of this Sacred Yoga

[2] This is how this Yoga was given to me when I was Arjuna. Before that when I was Dhruva I was given by The Word I AM the Gnosis of Sankhya which I propounded when I Was Kapil. At that time I also burnt to ashes 60000 ancestor of the Ikshvaku clan for their misconduct. When my Guru Krishna left, this Yoga was robbed from me by robbers of that clan supported by cosmic ruler Siva. Then they sunk the city Atlantis the Dwaraka established by Lord Krishna. Again and Again I gained my Yoga from God I AM and remade a new Dwaraka but they sunk that – by now six times, and every time they robbed my Yoga from my soul. In the war of this time – I understood the Mysteries of Great Yoga of God I AM and...the table is turned....

[3] New Testament names this Yoga as kingdom of God on earth. In one parable this is said to be like a precious pearl for which I actually sold everything else and this Yoga is ever and even more precious...

8. True Yoga is © Me

Gita 18.68

That who will tell this Supra Secret Yoga to my devotees after he has done supreme devotion unto Me[1] will undoubtedly attain Me the Word I AM

Gita 18.69

Among men on earth there is none who does unto Me a dearer service[2] nor shall anyone be dearer to Me[3]

9. Who Qualify for this Great Yoga?

Rev 21, 22

This Great Yoga, this precious pearl for which I sold out everything I had[4] is also the Water of Life of the New Testament

---

[1] In Rev 3.4 this is said as – I have loved him. The context is that the false Yoga masters of Philadelphia ka church = Synagogue e Satan will be told by Word I AM that He has loved me. This infers the zones of divine Bliss to which He led me

[2] For me this divine assignment is a sacred duty – in contrast for PM Modi this is merely a propaganda of selling what he has not nor has any divine sanction and this is a sin whose wages are…he must ask his advisors

[3] In this verse the phrase – neither anyone is nor shall be – clearly gives the exclusive © to me and neither to Modi nor US patent holder Vikram

[4] Elsewhere in Gita I was granted that I can reclaim My Yoga therefore I retrieved my Yoga from the robbers after I understood the Great Truths of Yoga of God I AM as He led me on the Way

The Rev 21, 22 have injunctions against those enemies of human values who do not qualify to receive this Yoga, further to that

Quran 7.40

Those who do not believe in Revelation Book will not enter the paradise[1]

Gita 18.67 says

This supra secret Yoga should not be given to a person who lacks dedication and drive[2], who is not a devotee[3], who does not want to listen and never to he who is jealous of God I AM and Word I AM

10. GITA AND YOGA OF INDIAN PM MODI

PM Narendra Modi does not have divine jurisdiction to teach Yoga of Gita not that he knows any. What he promotes are merely some postures but this is infringement and dilution of Glory of Yoga by making it cheap

---

[1] By another translation those who do not believe in the Signs of God do not enter paradise, the Sign of verbal equations and word Vorpal made from Verbal and Gnosis are some of His Signs...

[2] New Testament says – who thirsts for Water of Life

[3] How can they be devotees who do not know Who is God and Who Word? Well I told you that in this book, rest is now up to you. If you like I will tell you again when you Come if you Come in Trust

Even these postures – are based on sketches made by a Yogi supported by the King of Mysore. The Yogi had 2 disciples, one - famous as Iyengar was an expert on Pranayama and he also wrote a good book on Yoga postures. Another went to USA and established some schools which teach well. BUT the dynasty of the king of Mysore suffers from a curse even to this date – true Yoga could have cancelled that CURSE by goodwill and peace between the dynasty and a woman of past who had put this curse – even now the king may come to me and earnest and ask for a peace treaty and redemption

This is very mean of Modi to cheapen the name of Yoga in this way, worse – he is hopefully trampling upon my corpse, because Siva wanted me dead, even so mercy of God offers amnesty to Modi if he comes...to me

Gita 18.70, 71
These two verses of Gita permit everyone to read or listen to Gita with reverence – and get some divine rewards as specified. Promoting Yoga is not allowed. This is transgression...
Modi is promoting Yoga by Siva which Yoga was learnt by the snake in his neck who is Patanjali. My research shows that the Yoga given to this line was announced by Gita 4.2 to be lost – now waiting to end. Siva countered by robbing Yoga from Arjuna as soon as Krishna left, then he also sunk the city established by

Him[1] and pushed down soul of Arjuna. From below he again raised himself by Yoga of God I AM, again sabotaged, and thus Dwaraka was sunk 6 times by now. This time the Truth was exposed and Siva failed

MY RULING

All who knowingly transgress the Word of God...get wages of sin. Modi supported by UN is leading the world like the legendary piper led the lemmings

> *My ruling is that IF those who are so being misled learn this falsity and leave Modi and come to the side of True Word of God, they will not at all be considered sinners. Even Modi and other unauthorized Yoga teachers may repent and return and be saved*

---

[1] Atlantis, named in India as Dwaraka – that is the Door to the God I AM, therefore named Dwarka. By Bible – I AM the door

III. Exposing Fake Yoga

Yoga was used primarily as Martial Arts – a fairly bad misuse. First of these were developed in Kerala and also in China and Japan

The martial arts of Kerala belong to the immortal king[1] Mahabali and demon Vanasura who was his son. In this era he fought against Kalki and licked dust. The Yoga of Modi is an attempt to make a military corps of Yogis to fight against the God. One must note that in Yoga the mental field of a person makes a spiritual entity, all such entities following a guru join hands and they oppose and fight against other such entities and even humans, in this case they were programmed[2] against me and my kin

Try to understand this by phenomena of mass frenzy. In this also invisible mental entities combine to create a group entity of mass raze – more powerful effects are created by Yogis following a guru

---

[1] All immortals of past era were fake and are dead

[2] The word program can best be understood in context of movie Matrix, by that every manifestation is a PROGRAM of the Matrix and such programs interact in various ways, problem is caused by obsolete programs who refuse to be deleted and go in hiding. In present case Siva was secretly programming power of fake Yoga against God Eternal Son me and my kin, much of that was destroyed and Modi did a last phase effort which too shall die

YOGA AS PHYSICAL EXERCISE

Yoga used as martial arts was later converted in defensive use by Buddha – because the effects depends upon the guru who educates the minds of followers and also controls the invisible spirit effect program

As physical exercise what Modi teaches is Good for humans. But the inner intent of spirit directors is Satanic. When wrath of God descends the humans will be harmed simply because they followed a guru who is against God and serving the cause of Siva. The ONLY way out of this dilemma IS that one must NOT pursue the Modi or any other Yoga but come to the Right Hand of God

I know that gurus bind the followers in tight bondage which is very difficult to overcome. But Satanic influence is vanishing ady by day and yet if one finds this difficult let him or her pray to God that whatever is due from him to the guru that may be paid from his account by God and He may cut the bondage of captivity. In Bible this is told as – someone begged from God to give him more time to repay his dues and God granted, but there was someone who had taken a small loan from this man and this man pushed him to pay back and God pointed out the misbehavior..so pray to God and fight out if necessary but leave the fake christs and come to the Right Hand of God...or... Remember once you come to the Right Hand of God you are entitled to anything which is true good and beautiful if you have reached the stage to deserve that! Come!!!

IV. GREAT GOD AND GREAT YOGA

By Gita 4.1-3 one learns that Yoga is not a public commodity, but reserved for the rulers of the mankind. In Hindu Dharm the King used to have a Guru to guide and protect his kingdom. RSS BJP consider sinner Atal Behari as expert of Hindu Dharm but none of them have any True Guru...why?

The True doctrine continues. I shall offer Yoga and those who proceed shall on maturity become the rulers of cosmic arena[1]. There are 7 trillion life planets in this cosmos and they are organized and grouped in zones and subzones. One can read about this in the Urantia Book. Yoga is for the cosmic governance positions of these zones, subzones and planets in the new age. Now, one can see why Eternal Life is part of the offer!

---

[1] This may take several thousand years of Yoga. These timings and how much Yoga one gets in a life time and how many get the Yoga in different degrees – are all fixed by God. Yoga is a sacred Trust of God and not a cheap circus of Modi Ramdev Sri Sri etc.

GIFTS OF GOD

In the new age God I AM has much to offer to each human. These attainments enrich the life of persons and the society. Here is a non conclusive listing of some preliminary attainments

1. Liberation from Matrix
2. Bestowal of new Circuits of Mind and
3. Faculties of Mind
4. Initiations that is - Baptisms
5. Living fountains of water[1]
6. The Promised Land forever

1. LIBERATION FROM MATRIX

I recommend the readers to see the Matrix series of movies and make notes of the esoteric truths

- In era of Siva the mankind was forced to conform to Matrix. My research shows that on death the soul of persons of God I AM were robbed of the Yoga acquired by them which was added to their own kin. Matrix movie claims that such indoctrinated spirit-machines were using humans as biological batteries to provide life force to them and therefore the

---

[1] This is promise of Rev 8.17 and applies to all those who were led by Noam Chomsky and Susan Sarandon for No War Movement as is clearly seen in Rev 8.9. Now everyone is invited to the Right Hand of God

humans were growing in population as if being bred for such farming of life force

When the Book of John says that the Truth will set you free, the real meaning is in above context. The Matrix movie does not chew words of truths and clearly says that so long the humans are captives of the Matrix they are not free humans, let us note that they remain in the mental wilderness of I U and neither know the exploiter Satan nor the liberating God I AM. But when they are free from the Matrix – they learn their truth and start becoming HUMANS

This book exposes more mischief of Siva era. He made shrines of ashes of his first wife and they were used as spiritual batteries for his sycophant followers. These shrines are named Sakti Pithas and ironically one such centre is Mecca – where people revere this shrine but also throw stones on 3 towers which are symbols of the 3nity = Babylon whose Chairman was Siva. This is a depth of Satan who thus fooled Muslims. In fact Satan fooled Prophet Mohammad and he gave charge of Islam to two persons and after him two sects were formed, Stan entered again to create ruin and the two sects fight even to this date

Shrines of Siva are named as Jyoti Lingam. But there can never be a material temple of God I AM – the material tabernacle of God I AM was made by Moses, later a grand temple was made by my son Solomon, but both these were the earlier stages and both represented human body as to be developed in that era. In new age I AM the living Temple and if any material plane institutions are to me made they have to be for education and meditation

All powers of material temples of past era are fading away fast. Many such temples thrive on miracles - the very concept of a miracle is black magic hence downright foolish. Miracles are all innate in cosmos and are to be directed as Truths to fulfill the Purpose of God. Purpose of God IS to lead man to the Path, actually to the straight Path towards his ever increasing perefection, because man was made in His image and for this way and destiny. This perfection necessary involved liberation from powers of matter so that one can evolve towards Godliness. Liberation from matter requires one to pay the debts of matter – but sins burden them with more debts and thus mankind is trapped in chakra vyooh of Siva who hopes to show to God someday that He did an error by making humans. Media in India is now sufficiently educated to understand what I said here, even personal secretary of Modi is if not yet rest of RSS BJP and Modi and the Hindu outfits, let us see if they can 2 + 2 as 4 or not?

WHAT ARE THE TRUTHS OF HUMANS?

1. Only in human form one earns rewards for their actions. The reason is very simple. In human form there is veil of matter upon the eyes which does not let them see beyond the I U wilderness. In all other forms including gods there is no veil but they are as they are *made* and are not entitled to any rewards for what they merely ARE and did not strive for and attain. This is exactly why Siva feared that humans may surpass him, he also feared to come as human fearing that he will get lost behind the veil – he played too safe and too selfishly and lost

2. Siva is not the God, but he was only an executive with tenure. Spirituality cannot come by worship of Jyoti Lingam or Sakti Pitha far less other low status gods. Old era worked on model of Siva who had left God and wanted to show that he is right and God will be proved wrong. That did not happen. God I AM is not to be found outside but the Way starts Within[1] You – from your I AM[2] and I am the Guide and am setting up the cosmic system and new laws are being embedded in Time Space by God and me

---

[1] In New Testament this is told as – the Way is narrow
[2] That is the purpose of this book

2. BESTOWAL OF CIRCUITS OF MIND

There are 2 types of mind. Conscious Aware Experiential and Autonomous

> The evolution of conscious and experiential mind is based on 7 mind circuits. These circuits are like any other Integrated Circuits except that the material[1] is invisible and unknown to humans and within jurisdiction only of God and the techniques of His True Yoga

These are the 7 circuits. As far I know, the presiding deities of these circuits are changed and the circuits will decay in those who are on the Left Hand of God

1. INTUITION
Quick perception

2. UNDERSTANDING
Quick reasoning, conclusion and decision

3. KNOWLEDGE
This starts from curiosity – for true Gnosis, and couples with courage and counsel. For example either you reinvent the wheel or take counsel from one who is

---

[1] This material is known as Mahabhuta, named as earth, water, fire, air, space in India, differently in China but these names are Codes only

expert in making ball bearings. Here you can discover the importance of the Guru or Teacher as also dangers of false ones in best possible attires. If you assume that I must prove that I am better than Modi – Bible says that I shall not tempt you, but you have to develop Trust. Once again who to Trust? Bible says that – you must try to recognize my VOICE[1] or Come and use your logic – but I shall not tempt by super market techniques first developed by David Ogilvy

- *Here is a small Free lesson to you. David Ogilvy was not evil. He was sincere. See it this way – sword per se is neither good nor evil but can be used either way. Vorpal = a two edged (Rev 1.16) blade is efficient means for Eternal Son to do Will of God in two ways, one, protection of some, two, destruction of some*

4. COUNSEL

This is the social binder and by this even animals learn, did you see how a cat teaches her kitten? There are two aspects of this – by counsel alone the Gnosis of elders can pass on to the next generations, but if the elders get

---

[1] Do not try to recognize face to be like Jesus. Do not ask to see holes in my palms. Do not wish to first see me on cross. If you recognize my logic as resonant with some chord within you, you recognize my VOICE, then you must overcome the hurdles of Siva mafia and COME

stagnant[1], worse if they become deliberate enemies of the mankind, you do not have to imagine but can assess the present situation for these symptoms

> *Verily I again tell you that you enter the kingdom of God if you are like a child – BUT do not forget that a child is most vulnerable to cheating by Satan*
> *What is the solution?*
> *At least debate is not any solution?*

The solution[2] IS – I try my best to do Will of my Guru and God I AM and the Eternal Son who lead me and you try your best to seek True Counsel

5. COURAGE

I shall be very brief on this at this moment. Only courage can lead you ahead on panorama of human evolution – to keep moving in some circles is not my definition of courage but to stay at one station to fully learn and consummate and to counsel those behind you is also necessary

---

[1] This age was of long war of the evolving Eternal Son against Vanasura the disciple of Siva through who he wanted to excel. Devi Bhagwata says in the last chapter that Vanasura blocked the river by his 1000 hands. He was deliberately blocking the human evolution and black mailing God I AM for long. That is why end of 3nity = Babylon
[2] In my opinion

Essence of courage is to seek the next step upwards when you become set on a step of the Way. Mob initiative is not this type of courage – the evolution on the Way is primarily personal. Any lack of patience and intelligence which is natural in mobs always places courage in wrong direction. This was a major flaw in spiritual drive of the era of Siva in past about 125 years

6. WORSHIP

Animals are soul less. Groups of animals are controlled by a group circuit but that is not soul. No one has seen God and no one has seen soul. One learns about soul by self learning. Soul collects essence of your experiential learning in a life and transmigrates to next BODY who becomes another Person and NOT AT ALL YOU – you are mortal of one life. Soul is not yours, but IS using you and others for his progressive learning

- *Here is the significance of Eternal Life. If you come to the Right Hand of God I AM, you and your soul get married to each other and would even be resurrected when mature*
- *Soul connects with mind when a person is born. The consciousness is transferred from transmigrating soul to the mind. The mind makes his or her own concept of being a PERSON. At death the consciousness collects essence of learning and leaves. The person dies*

- *But soul can also be connected to God and that IS the way of mind and person towards God. This is where I stand to guide you*

- *This is the Key of Your Treasure. Read and reread to get the true perception. If soul is connected to God and your mind aligns with soul then your Person does not die, but becomes a companion of the soul and God. Resurrection is a process where God decides that your mind is worthy of becoming immortal and the marriage of Person and soul is consummated – in zone of Bliss*

- *Role of Guru is very important. When you enter zone of soul there are essences of learning of past birth and each birth has some unfulfilled desires which may catch you and pull back in time – Siva himself got pulled back in time to the curse of Genesis because he did not come to the Right Hand of God through me*

- *Most gurus of his era talked of kundalini but no one know what kundalini IS – this is the case of a blind leading the other blinds. Even I do not know what kundalini is because there is no need to know that – that was a name to cheat you. Some say that kundalini is name of power of Siva – then say so, why hide? Some say it sleeps and requires awakening, others say it is always awake. What I know are the Truths and names are not important. IF you are a seeker of Truth you recognize my Voice, if not - not*

Please remember that

1. Immortality can never be found by techniques of Yoga. The techniques are necessary but not enough. You get immortality IF God finds that by becoming immortal you can be of use to His Grand Cosmic Purpose

2. In the same way, Bliss can never be found by any technicality of Yoga. Bliss requires technicalities BUT you get bliss in proportion to the happiness which you have given or have potential to give to the cosmos, and the decision is made by God

Do not ask me the authenticity of above rules. Because I made these rules and I am a fellow of God I AM and the new age will have new rules embedded in the time space fabric by such Mysteries of God. I have not made anyone very happy but God gave me bliss zones because in His Opinion I exposed Satan and that would make cosmos happy, if not today then tomorrow onwards. God did not exactly tell me this – but this seems the only logical reason to me

Coming to the issue of Worship – this is quality of mind which can link mind with God I AM through soul

> *Worship was misused by Siva. Whoso worshipped him gets a boon. Ravana offered his 10 heads and became his greatest disciple*

Be Happy that God I AM would not ask for your head or heads, and THINK – He might ask for your heart[1]

Welcome to The Wonderland of God. Now there are no red queen, no red knight no Jabberwocky

7. WISDOM

By now you have become wiser. Either you have learnt something which you always wanted to but did not know what how and from who? Or you know that this book is not worth reading anymore and there are better ways to go about the business of life

I tell you two more mysteries

1. God I AM made puzzles in Bible. I opened the first puzzle about 16 years back – the 666 puzzle of Rev Book. The way is simple: you put a=1, b=2, c=3 on say both sides of Lamb = Om and check thus: 12+ 1+ 13+ 2 = 15 + 13. God made these puzzles because He IS Wise. Siva could not solve these puzzles but God made me do so when time of any puzzle came. Everyone knows of importance of codes in war – this is one example

God may give you circuits of Wisdom through Yoga – but True meanings are attained by using these circuits

---

[1] Caution – do not take this literally

in real life and by experiential learning[1]. And you become worthy of the kingdom of God when you learn how to use the Gnosis for His Purposes. This may sound like very self centered and selfish view of God but once you realize that all your True Good and Beautiful Wills are within the Will of God – you find the core of Wisdom about why at all to do Will of God

3. FACULTIES OF MIND

There are 7 faculties of mind at the present stage of human evolution. When you come to the Right Hand of God these are up graded for your success in life. I do not fully know but those on the left hand of God may lose these circuits, worse the personal presiding deities of the circuits may die right there and create complications...Coming back to the listing

1. LOGICAL

This is sometimes named Scientific Mathematical

2. MUSICAL

The name is self explanatory. However this faculty also trains you to seek and create Harmony in any desired field and this makes it very humane and important for family society nation world

---

[1] There is NO substitute of experiential learning

3. LINGUISTICS

Once you are in this zone, by and by you also learn that language is only a medium and not the Truth – but actually a barrier. And to know the Truth you have to necessarily deal with the barrier too[1] - more the practice and expertise more the refinement of thoughts and concepts. If you have aptitude do learn more than one language but Truth is elsewhere, not in language and yet Truth can never be approached without words and languages

4. SPATIAL IMAGINATION

It is said that James Watt saw the kettle and imagined stezm engine. Einstein did better – he did not see gravity but only knew that gravity strings pulled an apple on head of unsuspecting Newton...frankly speaking I do not know what all Einstein did but he too made some prophecies of the new age

Arts are 2 dimensional spatial imaginations, sculpture 3 dimensional. Einstein brought in fourth dimension, the time, and also invented things about behaviors of light. Today all students of advanced Physics know that

---

[1] My preferred definition of Truth is by Atothis in the Book Initiation by Elizabeth Haich. Says – Truth is like an invisible man, you can see the outlines if you dress the man with some cloths but you see only the cloths if you overdress him – by WORDS. I was Atothis!

light is responsive to human consciousness in certain situations and in mysterious ways

5. BODY AWARENESS

You have gravity responsive mechanisms in ear to know the orientation of your head. But more body awareness is used by expert surgeons, trapeze artists and even average players of cricket, hockey, football, volleyball, tennis and badminton

These are basic faculties. There are many other advanced versions. For example homing instinct of a pigeon may grow in humans as quest for God I AM

6. INTER PERSONAL

This is about of skills of communication and relations. This seems to be unimportant because one learns on his own even as a child. But this is very important. As one example can you see any issue with point of view of another, or many other persons? Or would you insist that everyone sees things your way? Dale Carnegie did not exactly say this – but gave a formula that makes the other person feel like a person and you get your work done through him. Before this could improve society, Modi got the idea of political use of this and the sin is on the head of the publisher of Dale Carnegie who should have gone deeper to discover more truths. Both David Ogilvy and Dale Carnegie pushed mankind in trap of flowery talks but without this how would one learn to overcome?

7. Intra Personal

This relates to I AMs within. In humans there are all I AMs from matter upwards to their own level of I AM from where the Way starts to That I AM. How to make them all healthy and interactive in proper way? There are very many secrets of Yoga in this zone. Self enquiry and guided meditations are main anchors of this most valuable faculty which is the key of the Treasure Chest of the human evolution

Matrix tells of a key maker who tells of many doors and many ways like New Testament tells of many mansions in the house of the father of Jesus. Second Book of Enoch revealed in Israel also tells of many keys – but that book interprets the keys in a different way - as musical keys of a piano. Matrix movie gives a better postulate. But what that key maker had were keys of the false and lower paradises. Matrix movie does not say that key maker was just another control point of Siva

The key maker led Neo to the Architect. Neo rejected the Architect. But Neo did not know that the Oracle and the key maker did not lead him to God – but to the very maker of the Matrix – men of Siva

Rest is simple. To know that the Architect was a fake IS the Key to find God – this is what completes the movie series

4. Initiations that is – Baptisms

Initiation is a term difficult to define. This requires a small particle given by the Guru to enter the mind of the disciple. In language of divine Alchemy this is described as – to make gold you need some gold to start with

> First - the bad news. One Indian legend says that a new demon was created from every drop of the blood of Raktbeej when it fell on ground
> - You see this in movie Matrix where Smith made his replicates in anyone he touches
> - In India this was seen by cap phenomena done by Anna Hazare

Now you know that if you blindly follow a Blind Guru worse a piper who sees where he is leading the lemmings but they remain blindly trustful to him – you are done with. When Bible asks you to come out of Babylon – this is implied that on your own you can not understand that this is Babylon and you must come out of that – think on this

SEED

In New Testament this is said by parable of Word of God as seed in different types of grounds of mind. Seed mean divine Gold. Bible tells of the different ground of mind and how the seed responds to the ground. The seed may be small in size but can be like an integrated circuit with huge information and decision points within. Bible tells about seed of Word of God and different types of grounds but does not say much about seed given by Satan – that is Water Baptism, to start with cleansing

Do not ask me why God does not tell plain truth. But imagine that if God would have told satan all truths on face of it – the Satan would have dropped atom bomb to destroy the mankind and to show to God his might

Time to become wiser; come out of Babylon and to the Right Hand of God

- The story of Adam Eve Serpent tells that seed of serpent in Eve made her lose immortality. Seed of Adam in a low plane woman did not harm him but he was cursed by God to go back to dust. In this story Eve fell because of scientific and technical reason and the other two by wrath of God because they failed His Human Evolution project, Serpent by mischief and Adam by carelessness. Of them Adam was cursed to go back to dust but serpent was cursed to lick dust forever. Adam was liberated but not the serpent – destined to fall in animal zone to assist in early evolution of life on new planets – that is the real meaning of being on the left hand of God, therefore Come!

LAMB

Lamb IS the Divine Gold Particle given to Mary Magdalene and you can know more about this mystery from the nursery rhyme – Mary had a little lamb. Lamb revealed the Rev Book to Mary but this was written under name of John. If you read Da Vinci Code you will understand this

Probably Mary was twin sister of John and she was secretly married to Jesus. She brought the seed – external covering of Jesus and iner core of Eternal Son so that someday this may be the locus of Maha Avatar of Narayana. Jesus was not Eternal Son but the human who did not live long enough to assimilate the truth of God – Jesus was murdered by his own father Siva who had earlier murdered his first son Original Ganesa too. In truth Siva never had a proper son. Ganesa was son only of Parvati. Murder was necessary to intimidate and to fit another head – so that such persons remain loyal to Siva

THOUGHT ADJUSTER

Urantia Book has several chapters about Thought Adjusters given by old era god Siva. The very name implies that he was controlling the thoughts in his own likeness – this was the biggest cosmic fraud known to me. To any child making a ethical decision Siva gave a Thought Adjuster to guide the child. Siva had given a command – be ye perfect even as I AM perfect and this particle was making anyone like Siva. The CATCH is that Siva was not at all perfect. At the most he was like Major Major of Catch 22

A careful reading shows that the command came from God I AM and Siva knowingly or unknowingly misunderstood this

Another mystery is that just like only one sperm enters the ova – only ONE man was allowed to have this privilege. God I AM chose me for this and before I could know what God gave me and how to develop that by technicalities of the True Yoga – Siva started my persecution in sheer jealousy, God stood behind me, I did not see Him and was frightened but Siva lost – I suppose Siva too did not see Him and that was the main reason that he lost. Now one sees that Siva was true to the model of a bakari and 3 monkeys of Gandhi – for love of wooden goat named de mo kurcy Siva would opt to be deaf blind and silent as if by oath of Omerta of the Italian mafia

NOW THE GOOD NEWS
BIBLICAL BAPTISMS
One baptism is given to the child as 3fold baptism in the name of Father Son and Holy Spirit. Jesus = a-z Ganesa was considered The Son – which is now not true. Now the God I AM is the Father, Virgin Mary the Kanya Kumari is the Son and Lord Kalki is the Holy

Spirit. These 3 are gods of the earlier 3 cosmic circuits of Personality Soul and Mind[1]. The management systems are under change, new officers will be appointed in due time, till then God I AM is managing the offices by default. There is no scope for jealousy because all old officers have reached the timer limit and are being made new. I add this on 25 August that yesterday an issue came up if the snakes were justified to oppose Arjuna because first he had attacked the Khandava Forest. The final Judgment says that Arjuna was doing Will of God and there is no appeal. The snakes agreed and many of them will enter the new age with new Life. Yesterday even Krishna was Judged. He cleared the issues but is being made new. Today is festival of Janmasthami in India and if I complete this book today I may add a note of second dedication – to Him. However all the future Worships go to Kal of Gita 11.43 as the Greatest Guru and I wrote that in the Section 7 of this book

WATER FIRE AND SPIRIT
The Biblical baptism defined in this way has direct True Meaning in the new age

---

[1] By the time of final editing of this book, Mary is promoted to Absolute zone, Kalki the CEO God number 1, Hatter 2 and Alice 3. By prophecy of Matrix both Trinity should have died and Neo followed her and also known as Savior. I changed the prophecy – Mary is alive and in Absolute Zone...

1. Water Baptism

This is similar to Indian custom of taking bath in sacred rivers. The scientific Initiation of the new age implies removal of past impurities

1. One such impurity is Jabberwocky the love god made from the ashes, and connecting the faculty of human love with the new age love God[1]. Yoga of Modi or anyone else cannot give this new connection – so must the humans come out of Babylon

2. Another problem is *reticulum* embedded at top of the spinal cord stem. This comes from reptile age and has fight or flight impulse

    ▪ Fight and flight can be a wise decision either for survival as developed in reptiles and carried over OR for power to assert Truth or to ignore the very importance of entanglement. In recent times this was demonstrated in USA in that the movement by Noam Chomsky and Susan Sarandon. They did assert truth but also knew that they should not do more than this. This event was foreseen in Rev 8.9 and all them have been chosen by God I AM for the new age destiny. They are the free men of the Rev 6.15

---

[1] Pradumna, a son of Lord Krishna

3. In general all the qualities of animals are to be removed, replaced and selected qualities upgraded, for example homing instinct of pigeon upgrades for humans seeking their own origin. This is where Modi utterly fails. He assumes that what we have on earth are the basic humans – but they are not, even he himself is now obsolete human and has in fact ben written off, yet may avail of The Mercy of God to repent and return

- We all originated in 3nity who are now written off, natural consequence is that all will be pulled to where they went. Their gold has been taken away and dross of the cauldron thrown in animal zone – that is eternal hell. To be saved from that natural destiny must one come out of the Babylon to the Right Hand of God

- The solution has two parts. First is to overcome this biological pull. Second is cleansing from – the worm. The cleansing part is Water Bath. Without this one does not qualify to move towards Godliness

2. Fire Baptism

Gita says that one can burn off their bondage by fire of True Gnosis. Another way is to cut off – by Sword of Gnosis. Both Sword and Fire mean same in this context.

In actual practice this involves cleaning the subconscious mind – the wilderness = a-z I U Babylon

This is most difficult part of Yoga. When one enters this zone he finds that his conscious mind has less power than the subconscious mind[1]. He gets entangled in past, to be precise in some unfulfilled strong wish of past life. This is like journey backwards in Time and if one is trapped in some past slot – he is done with. Prodigal son Siva was coming to God, but god blocked by Viswanath because Modi did not have enough intellect or faith or both to overcome that pull – so Siva fell in curse of going back to dust and now Modi has option to be on the same way or Come

Siva led Parvati to subconscious mind and showed her the skulls of her past births. This is the story of Amarnath – literally the Lord of Immortality. But neither Siva was immortal nor could he make Parvati immortal. This is known that Parvati left him and became Virgin Mary

---

[1] Like Dr Freud said – that mind is like an iceberg. That is to say the subconscious is 9 times more powerful than conscious. But actually the number is much high because subconscious mind is sum total of essence of knowledge of very many past lives. Men lived very many lives because Siva recycled them again anad again, never sending any one to paradise of God because he had left God, instead he made many mansions as his son Jesus told

Probably he wanted Parvati to become immortal and then rob her of the immortality – this is what he actually tried in an oblique way and God I AM showed him the clean mirror, yes – clean mirror: God welcomed the prodigal son back to Him

Amarnath is a great shrine for Hindus. Why not? Everyone wants to be immortal. But they are all under temptation – because Siva failed and he could not make Parvati immortal nor he intended to make anyone immortal

It is said that a mantra given by him wards off death. It does not. It is said that the secret of mantra was given only to Parvati, even Parvati was not made immortal. All mantras of Siva are like promise of Achhey Din by Modi. If one goes by Words of Deut 13 – Modi is finished because he made a false prophecy in name of Siva, Modi may not see this **writing on the front wall** only if he is totally blind

Siva gave Mritunjaya mantra to defeat death. It is very sacred for Hindus. So far no one has escaped death – but they all remain tempted. Modi was first politician to overly incite temptation in needy Indians and knows nothing of the practical side of how to do what he talks about. In both cases salesman ship

Because of the difficulty of this Baptism – a True Guru is required for Yoga. Only Gita tells the reality in the verse where Krishna tells Arjuna that He shall absolve him of all sins of the past and he must not grieve. Only God who spoke through Krishna can give a clean slate of subconscious mind – that exactly is the Promise of the John 3.16

Can Modi do this? Siva was in him but Siva got stuck to his past in Varanasi and other places. God gave clear offer to prodigal son to come to Him, but he lost full two years before failing...now Pope would know that ritualistic baptism is not reality but only a memorandum. In the same way Muslims must know that Vazu by water or even by sand if water is not available in a desert is actually a memorandum of Water Baptism...no more. At the time of final editing of this book on 25 August 2016 we all see severe flood in Varanasi. News reports say that Modi is worried about this – is he really concerned? Then come – there is no other way. God shalt not change His Word

3. Baptism by the Spirit
This is the Gold from God I AM. I attained that after past about 80000 years. In India this is named becoming Dvij or twice born. Many Christians claim that they are twice born – but they say this on basis of some para normal experience. One can have several thousand para normal experiences on way of Yoga, the more incompetent the Guru more such traps – but these mean nothing so far True Yoga is concerned

God I AM has already marked who get how much Yoga and how they proceed and when they attain goals on the way – this is between me and God and actually God has written clues in secret puzzles, so far sincere believers are concerned the most important factor is to first find the True Christ and then TO START and be on the go...once on the Way you have actually WON and there is no further need to envy those who are ahead of you. The truth is that is you then pause to help those behind you actually get ahead without this being visible in physical terms. If Modi has any eyes he can see that I stay for this reason only. God Has given me His blank cheque book but I want to loot His bank

Equal Opportunity Offer

Though I have used Bible terminology here, the offer of Yoga is between God I AM and the human/s and entirely independent of any religion. The basic rules are that One does not worship any one of past or present but only the God I AM and/or His First Manifestation known as Narayana in India whose final and tenth Great Incarnation is personalized[1] in Lord Kalki who was born on earth about 18 years back and lifted up as in Rev 12.5, I Kings 8..19, Isa 7.14, 66.7,

---

[1] So far the mankind does not know the meanings of Field and Person who personalize *that* field. They do not even understand that mortal frame does not become divine by incarnation but only after the human mind cuts across the veil and discovers truths, imbibes them and proves in practice. But these are NOT any problems – these are the Ground Conditions of this Game!

Luke 1.35. Second rule is that Free Lunches are now not available even in Google. You receive something of value from above and also give a return[1] plus pass on something below you too. By rule of Devi Bhagwata the cosmic ruler pays 1/16 to God I AM, the rule cascades. To start with I ask for Tithe of Righteousness...from you at the top of the nations or money

If the personal secretary of Modi can find above rule in 5th Section of Devi Bhagwata, he may per chance also figure out that Siva the prodigal son failed because he left home and was not paying the 1/16 part to God. Actually he could never get that Yoga by whose Yagya he could give anything to God. He cooked up a tale that he was expelled from Yagya therefore he also could not give any to God but Siva was getting his share from the conspirators who made this tale

Instead of offerings to God through me, like Mormon Bible says the VIPs of India have robbed God and even the whole nation is cursed. When you compare me with Modi – I bring the news of India being cursed and Modi talks of bullet train to show God greatness of India of Modi...Modi is better salesman than I would ever be

---

[1] This is told in one poem of Gitanjali by Tagore that when God came to him, first He said – what you have to give Me!

I am inclined to give a small free counsel to Modi not that the price of the counsel is that. Jawaharlal Nehru was a big sinner. When I had encounter with his soul he claimed to be a Sadasiva who ruled over 36 more Sadasiva rank deities. At that time I was a starter and belived him, later I did what God taught me long back. Even Nehru knew that things will get bad. He clearly said in Discovery in India that – causes have been made and the train of effects will be coming. Modi is making nonstop causes and bringing better trains from Japan and thinks not of the train of effects which has probably come silently on his platform

Another example may be noted by Indu Jain of TOI. Nehru tells of a story of South India where a woman – most probably Jain – was robbed off her special gold bangle filled with rare pearls. She actually became a victim of greedy misunderstanding by a rich queen who thought that this was her lost bangle because richest of citizens could not have afforded that type of precious bangle. This did not occur to her that both bangles were made by the same goldsmith and looked similar The kingdom of Mazura was destroyed by this sin! I did not go deep in this story. Meenakshi the princess of Mazura wanted to marry Siva and above woman of that bangle must be in that lineage with a divine bangle

Listen Modi, if at all Ramdev was a true guru he would NOT come to you, but require you to come to him.

Hindu Dharma has a true protocol and Atal Behari is the last person who knows anything about Hindu Dharma, I told him but he does not introspect...but without fire baptism, without cleansing past sins embedded in soul – you do not ride white horse on Way of God I AM. You ride an ass, worse a wooden goat

As on now the sin of sufferings of all Indians is on your head, you deliberately caused this to them and even I do not have any idea how much hell you would face for this. All that IS needed of you IS that you read the parable of the prodigal son fully and carefully...you can not do that and claim to be savior of the world...is very funny even though you look serious

5. LIVING FOUNTAINS OF WATER[1]

Living fountains of water infer the Promised Land. This book is about the Way to this Promised Land – when you step out and is not academic exercise on papers of this book

Here I would add a note about Noam Chomsky who made a theory that human evolution is proportional to evolution of language. This is a true idea BUT actually the human evolution starts from thoughts, and then

---

[1] This phrase from Rev 8.17 relates to the Rev 8.9 people who are of the No War Movement by Noam Chomsky and Susan Sarandon

comes to words and language – but in reality the new age evolution comes when the True Words are manifested in the actions of a man... fisrt the mind is cleansed then tuned to doing His Will then the Word comes then Thought is formed then Action – that is Yoga. Modi has none of this yet he sells Yoga on streets of the whole world – what profits would that make if you lost your soul to Satan? Here is my Call to Noam Chomsky and Susan Sarandron not that even Noam Chomsky can fully understand what I said in the above lines to upgrade his theory of human evolution. May I too BOAST Sir Noam Chomsky that God Soul Yoga can never be explained in words – now that I said this, even I think that it is True and I did not know this before this time, Thank you that I remembered you and learnt more from God

Bible has a clue that Noam Chomskey was once Jonah the Yunus and Susan Sarandon was Sheba married to my son The all time Great King who did not ask for gold from God but for understanding of human nature, God of course gives something more when he likes a guy[1]...Solomon made the Temple and God gave me His Own Living Temple – I suppose if I were God I will be bankrupt in a week...that now I am bankrupt is by courtesy Satanism of my nation India, Modi?

---

[1] I looted God after I learnt this. I was told that I cannot find a big enough bag to store what God gives – so I asked for the bag too

V. THE PROMISED LAND

There are two meanings of the Promised Land

1. A human body with True Divine Yoga. Siva and clan supported by him lost this Yoga as told in Gita 4.2
2. A land or nation for those who follow the chosen one. Siva wanted this land and started drive from Ganespuri in India reaching to New South Fallsberg in USA and branches worldwide. He also created several gurus who are now exposed to be false Christs and powers of all these vanish as told in the last chapter of the Book of Daniel
3. The most vital factor in this context IS that the Promised Land was rejected by the Jews in times of Samuel and when I Sam 8.22 says – Every man go to his city, they were ordained to go back to their 'origin, the animal zone. The curse with all other curses is now open and so are open the blessings IF they now listen to me
4. Modi who seems to be the leader of the world is the main hurdle on the way of God – mankind must reject him and select the Way of God as I tell - IS the only way of Eternal Life and the Promised Land

SIGNIFICANCE OF THE PROMISED LAND

This is Yoga which was not there at that time therefore this was Promised for future. The Promise is for them who thirst and who waited and who now LISTEN and who COME...Bible says that Modi gets only dew on grass to lick, which is his Yoga, nothing more

PRODIGAL SON

Siva was the prodigal son. New Testament says that he begged from God that he will pay his dues, so God sacrificed fatted calf for a feast to him as he started to come...this he did through Modi

> *Siva failed. He could not come to God but got stuck at Visvanath and Jagannath temples if not in movement for Ram temple. I do not know if Siva did not deliberately come to God or veil of ignorance of Modi prevented him – but he failed and now it is easy for Modi to RISE beyond his restraining and in fact deadly influence*

Modi chose a personal secretary associated with Vivekananda and made special efforts for this. Now I see that this was done because Siva told some human in India that Kalki was first Ram then Krishna and then Vivekananda

This is true but the above last incarnation was superficial. Vivekananda was under influence of fake guru Ramakrishna and Patanjali[1] doctrine

Vivekananda meditated at Rock in South India Sea and there Virgin Mary the Kanya Kumari contacted him and asked if he will do the Will of God. He said Yes. She told that the time for him will come after 100 years. So he died young

- *In his typical management way Siva took control of ashes of Vivekananda[2] and created some entity who entered some human and claims to be Kalki incarnation in India...and there are more than one claimants to add to the confusion*
- *Modi too is trapped in a fake perception that Vivekaknada is assisting him*
- *But Modi may be surprised to learn that Siva played this ash trick even on Krishna and created Vitthal*

Siva failed in rational assessment and my Judgment. He entered a priest of a Laxmi Narayana temple and accosted me on a street and tried to involve me in flowery talks to fool me. God spoke through me and countered many of his tricks. In one he told me that he is my relative – I answered that Krishna asked me to slay

---

[1] The snake in the neck of Siva

[2] One may note that after death of Moses attempts were made to take control of his ashes for the same reason

Bhishma who was my grandfather. He said that since I became a warrior by fighting so the enemy served a purpose useful to me – I replied that they did not attack me with that purpose in mind. I told that my kin were murdered by Satanic back stabbing - but they rose very high in eyes of God, at this the cunning guy asserted that therefore I have forgiven them all – I told that what matters is if the innate cosmic law forgives him. This was almost a full one hour talk in the street; I did not know that Siva and his corps were there and that they were being Judged by God speaking through me. I realized the truth after contemplations for next 2 or 3 days in context of several interlinked events

KIND ATTENTION - MODI

India is the Promised Land – neither Israel nor USA. In Israel Siva had put up his claim by Al Aqsa mosque even as Mecca is a Siva lingam. In India Siva entered you to deceive God and in USA Ganesa entered Obama for the same purpose

Bible hints that IF India fully fails the privilege may be given to another nation – IF Modi has even one billionth part of love of as much he tells about India, he must note that he is skating on very thin ice of false perceptions. Love of land is OK but has no value if he does not love God

NOW PLEASE SEE THE MORMON BIBLE

Ether 1 tells story of Jared and his brother who fled together after fall of tower of Babel. One of them was a large and mighty man and a man highly favored of the Lord[1]. They called for Mercy of God and were answered and granted privileges. They were asked to come to the Promised Land...that is why we are here, that is why Hindus are here...

- *Ether 1.42 - God Saith: ...And there I will I meet thee, and I will go before thee into a land which is choice above all the lands of the earth...*

- *Ether 1.43: And there will I bless thee and thy seed[2]and raise up unto me of thy seed, and of the seed of thy brother[3], and they who shall go with thee[4], a great nation. And there shall be none greater than the nation which I will raise up unto me of thy seed, upon all the face of the earth. And this I will do unto thee because this long time ye have cried unto me*

---

[1] Ether 1.14 – There are many possible combinations. Rahul and Digvijaya Singh, Modi and Amit Shah are two – but in both cases Rahul and Modi do not have any lineage therefore the prophecy prima facie does not fit the requirement of Ether 1.43 on them and they must quit the race. One possible combination is my father and Ras Behari Bose who had met him, another may be my father and Jamshedji Tata

[2] You do not have lineage but followers...

[3] Seed of any and all second brother/s who come to God

[4] They are the Hindus of India...

Siva wanted his own persons to get this privilege. First he entered me so that he learns all the secrets of Yoga, but God asked me to request him to go out while the Temple is made in me. I made the temple for many years. Made and remade and tested. Then I told God to do this while I serve – that was the Master Stroke!

Then Siva entered you and he also ensured that you do not have a seed so that Siva can choose his persons to have the secret while you go to ashes. Thus Siva responded to the call of God for the prodigal son to come to Him

Siva fails, you may move in same line by momentum but you reach nowhere. Remember that once you became Great VIP in this world and now miserable before problems of this nation. You can OVERCOME now…do this… it is easy for a camel to pass through the eye of a needle…try…come to me…I would help…I have divine jurisdiction…and mandate to help

PLEASE PAY ATTENTION
Love and anger of God is SAME
Vorpal = a two edged (Rev 1.16) sword. This sword saves and also destroys

YOU HAVE FAILED MODI
You have failed to come to God as told in the parable of the prodigal son. The parable does not set any time

limit for you to reach God – He favored you when He saw you coming and set no precondition of time limit...but you have failed by your own words...see

MATT 12.37

For by your words you will be justified, and by your words you will be condemned

1. Whatever you promised to do, you will be required to do and then will you be justified – but you only talk and do nothing. This is first failure

2. *You said in context of your ministers that 2 years is sufficient time for performance appraisal – by these words you set your time limit Modi and you fail and STAND as desolate...by prophecy of 70 years cycle in the Book of Daniel – the consummation will be poured on you...double of the sin of the Mother of Harlots of Rev 17...that may be horrible when the time comes...therefore think and Come...so far I know there is no other Way for you*

I cannot overrule God but in my role as Savior I can ask for His Mercy if you can hear see understand and act on His truths...God says you come through me and contact me by 30 September 2016, and you *don't move* in any other way because the Damocles' sword hangs on your head by a very thin hair

DO NOT FAIL MODI, NOT ANY MORE MOMENTS

Read the decrees of God Modi and talk less think more...do not fail Modi, not any more moments because the sword hangs by thread of Time which is a very thin thread and keeps clicking too

1. *Ether 2.7: ...the land of promise, which was choice above all other lands, which the Lord God had preserved for a righteous people[1]*
2. *Ether 2.8: ...whoso should possess this land of promise, from that time[2] henceforth and forever, should serve Him, the true and only God[3], or they should be swept off when the fullness of his wrath[4] should come upon them*
3. *Ether 2.9: ...behold the decrees of God concerning this land that it is a land of promise; and whoso nation shall possess it shall serve God, or they shall be swept off when the fullness of his wrath shall come upon them. And the fullness of wrath cometh upon them when they are ripened in iniquity[5]*

---

[1] Are you anywhere near righteousness?

[2] The TIME starts when I declare the meaning in my mind - now

[3] How many gods have you served recently Modi?

[4] Fullness of His wrath comes within 70 years cycle. You plan 70 years festival of Yad karo Kurbani and sword hangs on your head. Bible says that the days will shortened for the sake of elect and I HIS ELECT say that the Day is Today and last date for you to start coming towards God is 30 September 2016

[5] Presidents, Judges, politicians, media – all have cheated me the apostle of the God in Constitutionally guaranteed fundamental rights and how come you play your band worldwide - band of hollow blab la?

4. **Ether 2.10** - *For behold, this is the land which is choice above all other lands; wherefore he that doth possess it shall serve God or shall be swept off;* **for it is the everlasting decree of God.** *And it is not until the fullness of inequity among the children of the land[1] that they are swept off*

5. **Ether 2.11** - *And this cometh upon you O ye Gentiles[2] that you may know the decrees of God—that you may repent[3] and not continue in your iniquities...*

OPEN YOUR EARS AND LISTEN MODI

1. 3 Nephi 24.8: Will a man rob God? Yet ye have robbed me. Yet you say: Wherein have we robbed thee? In tithes and offerings?

2. **3 Ether 24.9: Ye are cursed[4] with a curse,** *for you have robbed me, even this whole nation[5]*

3. *Modi I heard your main points on 15 August 2016. You say that you checked inflation to 6 instead of earlier 10% despite famine for 2 years. What you did not say is that by doctrine of*

---

[1] Thus the whole nation is cursed

[2] You were Jews earlier, now fallen and became Gentiles

[3] You have to repent first and then proceed, I will set time limits for that too, first for redress of Injustice to me

[4] You are cursed Modi, India 666 is cursed; whole nation is cursed – only if you knew what curse of God means, even I am shaken...

[5] Whole of India is cursed and you see the wrath of nature and Modi the total sin is on your head and of Atal Behari Vajpeyi – your IDOL

True Hindu Dharma the famine indicates your sins – and the sins are chasing you, your best hope was to have gone to Olympics for race around the world so that you can claim all the land what you run across before the sins catch you up – then would you find how much earth you actually deserve and need[1]

PLEASE LISTEN MODI

Caiaphas the Head of religious Sanhadrin murdered Jesus to save his religion and nation from Truths of God!!! He became Atal Behari and using the same words he rejected my call for Justice. DPM Devi Lal was inclined to impeach the Judges and VPS would have done that but BJP blocked this. That is why VPS who was Vibhishana could not come to God and died with cancer. Probably you think that a wise God would support you rather than me – but God turns out to be WISER than you, so you stay under the hanging sword like a big fool. If you cannot do anything about my Justice about which you took oath on Constitution why do truckload of blah blah that you would do that and that too please God?

---

[1] In one poem I had told George Bush that he can get enough cubic feet land if he multiplies his height length and width. The context was that I did not want him to start a world war – he not only conceded but also replied in public that it was my mere thought that he would start the world war...now I find that he was once Joseph and must come to me

Magic by Modi

Isa 1 says that Judges and Ministers must discuss with me so that their sins are made white by God. By what magic Modi you pulled Judges and CMs to your conferences and CJI gave you his precise Accountability that first you give him 70000 more Judges than will he take up my Case of Rights? Bible is clear – IF CJI and the Judges RISE they can make the new age history of India

My Case of Rights

My livelihood was robbed when as a selection grade confirmed officer in HAL posted in Lucknow, I was summarily dismissed after I send a complaint of gross mismanagement to the Chairman Air Commodore Kurupad at Bangalore. The dismissal was prima facie Unconstitutional and mala fide

Soon thereafter Emergency was imposed in India and my right to seek relief from Court became controversial and CJI P N Bhagwati back stabbed the Constitution

> *Now I know that Siva got me dismissed because he was suspicious of anyone finding faults with his systems, he feared that such a person may eventually expose him*

Then the Satanic system attacked my kin. After end of Emergency I approached Court – and once again the otherwise all time great CJI P N Bhagwati back stabbed me and after that very many Judges did the same. Later Indian Express was garroted because they published that SCI erred in at least 5 cases – and my case was one of these. Siva took hold of mind of Arun Shourie and he backed out and put his force in Break the Ayodhya Structure issue. However he was thrown to dogs by Goenka in silent connivance of the Fishy Judges of the SCI. Later, having noted all this Times of India decided to push me off the precipice and to please the SCI they sought a certificate from them that TOI was a very good newspaper – this is not a Constitutional duty of SCI to issue such certificates but they entered the *Banders Scratch Backs of each other* Mode. Complete national media secretly boycotts me even today and the invisible system of Siva kept attacking me and my kin and made me fully desolate. God gave me privileges of True Yoga and this further infuriated Siva. He wanted to remain at top of cosmic ruler ship and his man to become new age Maha Avatar Kalki. I did not know that God has chosen me and Siva kept on persecuting me and my kin even when I was his earnest devotee. Then God led me to The Truth step by step...Siva started losing the ground and the end came...

GOD HATH FORESEEN THESE

God had foreseen what they would do to me and had passed the Judgment beforehand in Bhagwata and the sections of Uddhava Gita. Siva did not implement these but now he is not the ruler – do you see the point PM Modi, CJI Thakur, Prez Mukerji and TOI Indu Jain...etc!

A. *He who robs the livelihood – whether granted by him or others – of a deity or Brahmin, is born as a dirty maggot for millions of years[1]*

B. *Same fate that awaits this culprit in the next life also applies to those who aid instigate or abet him in this deed or are sharers of it. The punishment is greater in proportion to the enormity of the sin*

ı

UN SPONSORED MODI YOGA
ALL NATIONS COME OUT OF BABYLON

This Yoga is exposed as stolen, is unauthorized, was misused and will face wrath of God. Right now God has extended mercy to Modi and given him time to consider the truths. All nations must come out of Babylon and be saved IS my suggestion

---

[1] Bhagwata 11: 27: 54 Uddhava Gita XXII 54

VI. Rules of the Promised Land

There are 2 types of the Promised Land. One is a body, mind, soul on Way of Evolution. Another a land or nation

Wonderland of Alice exists on subtle plane. There a Jabberwocky was slain. Later the source of the field of Jabberwocky. It was long war – demons were made alive again and again and slain. This love god made from ashes lives in all humans on earth plane – now that the Core Demon is gone all others will die a slow death. Men will become loveless. This is one reason why you must come to the Right Hand of God because then will your love be connected with new age field of new god of love. There are many other gods to sustain the circuits of body parts and this IS true for all of them

The Jews

Souls transmigrate. The Jews who crucified Jesus were persecuted by Gabriel the eldest son of Michael and he came as Hitler to do this. Now those souls are in RSS, they imitate Hitler salutation in a somewhat different way so that they too might be the winners and not the losers – by this bandergiri

On the other hand the Jews of Israel are different. They have superior wisdom and their ancestors showed drive and courage to come to Israel – in doing so they suffered and some of this I read in the novel named

Exodus. As yet neither are they in Promised Land not they have the True Yoga. What they have are 12 fountains which are cranial nerves – in fact all film actors have superior cranial nerves because some of these control the facial expressions. Even the stones of the river Jordan, the construction of Moses tabernacle and Solomon's Temple are secrets of the then Yoga – which is now advanced beyond these

The New Jerusalem is actually the new set of nerves and infrastructure of body and brain even of thoughts and soul – that come by True Yoga. Jews have been invited to come under Laodiceans = One Lord and are required to HEAR

BASIC RULES

1. Righteousness is the start up rule – in this Modi and India fail because of conspiracy against Justice. The whole world must rise against that and NOT rise to the temptations by Modi. Those who fail here cannot be accepted on the Right Hand of God and must pack up for the everlasting Hell. Like Nehru said in Discovery of India that causes including conspiratorial tryst with destiny have been made and train of effects will be coming

2. The next logical rule is that Promised Land is for those who serve God – the One God whose name is

I AM and this is clearly told by the Mormon Bible Ether 1

3. But the ground rule IS that by I Sam 8.22 – the Promised Land has been taken away from the whole mankind who are cursed to GO BACK TO THEIR CITY – the city from where they came, the animal origin. Only God can reverse That Status!

4. Jews contradict New Testament because their Torah says that hell can NOT be everlasting but should have a fixed tenure. They are wrong – the everlasting hell means fall for ever in animal zone as clearly implied by God speaking through Samuel – and the bell is ringing, so the *cat must be coming*

WHAT NEXT?

God is not a bag of chocolates, at least not my bag. He Giveth a clue and expects me to strive for truths. Thus I do not know what is next but get instructions step by step and this is clearly written in Bibles. In earlier days this was different by rules of Siva – the apostles were given precise instructions. But there was a negative side of that – they did not earn as much I do

With the information that I have, we cannot say that Promised Land will be a specific nation. But this land means a group of persons in various nations who join the True Yoga and proceed on that Way. Even the number of such persons and degrees of Yoga has been fixed by God and clues given in Bible

Modi cannot complain – I gave him many chances, God gave more. Now the last date has been fixed. If Sonia and Modi expect God to answer directly to them, please note that this is not legal – in days of Moses you have said that you do not wish direct dialogue with God, this is the present situation which no one can change...you either listen to me who am the apostle of Deut 18 or continue in wilderness = a-z IU Babylon without Him

VII. THE 70 YEARS PUZZLE OF THE BOOK OF DANIEL 8.26

Today the 5 August 2016, God I AM revealed the complicity of Ardhnariswara to me and that cleared the complete picture before me. The matter had past karmas which resurface, people die to become ash, ash smeared by Siva on his body and again given life for next round and so on. The Adisakti was created by Ardhnariswara who created 3nity and Siva burnt to ashes the Adisakti after taking her 3rd eye. Later the Adisakti was made alive as Nirmala Devi who used to put a very big red vermilion sign on her forehead as reminder of her 3rd eye. As matters stand today all of them are gone – all things are being made new. If Modi and Obama rely on these mighty deities the time for their disappointment has come

Sometimes I feel shockingly amazed at dare devilry of persons like Sonia M M Singh Atal Modi Obama who know that they are in dark within but who switch on lime lights upon and around them – for at the most one life, which is not more than a bubble on the endless stream of cosmic eternity. Sometimes I recall Einstein that world will end not by devils but by bystanders who do nothing to save the situation – specially the Journalists. They have the potential to rise but they serve under devils...If Judges of India RISE the table gets turned...but they are too weak as yet...only a WEAK Judiciary could have punished Aurundhati Roy is a clear proof

Freemasons were of lineage of Hiram the architect of Solomon who made the Temple. They came to UK and made a great Judiciary without any written Constitution. Some say that their Justice was based on their tools of the trade – spirit level for equality, Square Angle of square dealing and plumb for uprightness. Indian Judiciary fails despite truckload of Greek and Roman documents or may be because of them

END OF ERA OF VANASURA

PM Modi is very proud of Indian heritage. This story is dedicated to his grand follies. Vanasura defeated Kartikeya the eldest son of Siva and commander in chief of the *devatas*. He surrendered the flag. Parvati said that when the flag falls that will be the sign that kingdom of Vanasura will also fall. Probably Parvati had made up her mind to leave Siva and be born as Virgin Mary to slay Vanasura

Flag of tryst[1] with destiny was unfurled on Red Fort in Delhi on midnight of 15 August 1947. The same day Nityananda of Ganespuri gave his *charan padukas* to Muktananda who kept them on his head and danced. By the morning the flag of silk protein was eaten up by the crows – the Red Alert was on

Here comes the prophecy of 70 weeks in the Book of Daniel, to be exact prophecy of 70 years. This will be completed on 14 August 2017 except that the days will be shortened for the sake of elect. This means that dynasty of Vanasura may end any day. I told God that so far I am concerned the end may come on 16 August 2016. (I wrote the last page of this book on 26 August 2016 and that seems to be the clear Sign of End of a cycler)

5 August 2016

---

[1] Sort of pleasing conspiracy – of Siva and Vanasura

❖ YEHVAH CHRIST ALLAH; REV 6.15 AND 7 CHURCHES

Today the 6 August 2016 – God told me secret of the word Allah, and then of Yehvah and Christ. I explain in that very sequence

1. ALLAH = I 2 I AM

By this definition – Allah is the code name of the straight path[1]. The path starts from your own I that is the Khudi and by True Yoga leads towards God I AM, in other words from Khudi to Khuda

> *In contrast the path told by Siva as Guru was to meditate on him or his assigned gurus or their statues*
> *The difference is not of worship of statue and worship of Allah BUT the very worship starts*

---

[1] Chapter 1 of Quran, Al Fateha – Guide us to the straight path, this is the Job assigned by Him to me, and this guidance is of the Way of the True Yoga

within YOU and towards Him – as guided by His appointee

I am that appointee Ahmad = a-z. There are 3 requirements when I can successfully do the divine Job assigned to me

1. *When the Time is ripe and God gives OK*
2. *When I am ready*
3. *When you are ready*

The first 2 are OK – onus is upon YOU

2. YEHVAH = I --> 1 I, A-Z I AM

This is the Original Name given to Moses and this has a very deep definition. BY this – Yehvah IS the path of your personal I evolving-->towards One I the Who One I IS also the a-z Omnipresent I AM Whose a-z Existence and Contents are not known to us. And this evolution is by True Yoga

3. CHRIST = I AM IS Z

The definition given to Jews is very deep and profound. That was not understood; even I cracked the code after many trials

But the code for Christians is simpler – God I AM is final I AM the Z I AM

All these 3 sacred names of Bible and Quran are the Most relevant and presented this way they indicate that Way of Life starts from the I AM of the person towards That God I AM – and that is True Yoga and this is the Great Mystery of the True Yoga and I am the God appointed official teacher – Swasti

Please note that name of God is not relevant but is considered valuable because men were using the name and books for personal advantage and magical practices. Above equations indicate that the True Worship of God and Way of Human Evolution starts from the very consciousness of the person and depends upon his inner sincerity and not on outward rituals – Guru or guide in most important for Yoga and the God appointed person cannot be bypassed or imitated, not any more...

In contrast Siva would never have told you this – because he as the prodigal son left God and wanted that only he be worshipped. Therefore he gave path of repeated chant of his mantra and meditation on statue of guru

Purpose of this Book

Is to save the men of Rev 6.15 by calling them to the right hand of the God I AM

The men of Rev 6.15 were spirits who were hiding in rocks and mountains also within human bodies within caves of body and mind and rocks of bones. They hid because of the wrath of the Son – and they were forced out and thrown in fires

NOW this book invites those on the human plane in who these spirits had entered. The clear example is that the US Congress gave 7 standing ovations to a mere flowery talk speech of Modi and this very Congress had banned his entry in USA. This is a clear case of mass hypnotism by Siva who worked through Modi. The US men are damned by Deut 13. Their minds close eyes to this but souls know and therefore they welcomed Modi as savior – but he is a faker who used the fear of souls as locus for his black magic

These men are:

1. KINGS OF THE NATIONS

The kings of the earth and they must come with the honor and glory of their nations – that is, with the ministers and Judges, and also intercede for them that they will follow the directives of the God I AM through my office...this is true for who opt to come. Modi may come but unless God accepts him the other kings must keep him at distance lest they too may be rejected by God. Same about Obama and others

There is a clear message in Bible that the kings can not persecute their subject – but impliedly must intercede for their Judges and Ministers and then follow the guidelines of God through the apostle

The temples and trends of past are over and Quran clearly says of such provision of TENURE

2. *THE GREAT MEN*

Like Pranab Mukerji Modi Obama CJI Thakur Sonia Rahul M M Singh Queen of UK the heads of 5 Veto nations – Modi is greatest of all because he got 2.21 crore followers on twitter!

3. *THE RICH MEN*

Like those listed in Forbes Fortune. I am keen about Laodiceans = One Lord Jews, Bill Gates of Pergamos = Buddhist G

The rich men cannot go to paradise is clear dictate of New Testament...but God can overrule this...if they come to the Right Hand of God

This is linked with mysterious camel pass eye needle puzzle and also appears in Quran. I found several meanings of the camel puzzle but the most relevant is that where camel G = Modi and this is explained elsewhere in this book

The culture of Kuber as Rich Men was created by Siva. His idea was that so long the wealth of nation is in hands of his men – God cannot do anything to remove him from the chair. But Siva failed!

4. THE COMMANDERS

Like Kartikeya and others of the corps of the Siva of now fallen 3nity = Babylon including Babylon = U snake. These spirit plane commanders have their links on human planes. Putin is one – a trained KGB commander like James Bond. General V K Singh of BJP who might have tried to remove Congress Govt by armed rebellion is my preferred man – RSS corps too falls in this class

On 27 August 2016 I saw a TV report by IBN7 about Jaspur in Chahttisgarh – about snakes. Snakes too had commanders of Siva and he had given them a special gem which could make dead persons alive

Kundalini Yoga of Siva WAS actually about Snake Power. This is very clearly told by a Calcutta High Court Judge of the times of Britons who wrote about Indian Tantra by nick name of Arthur Avalon

On the time of writing this paragraph the issue of the snakes has been fully Judged and the human leaders in that class can come fearlessly to be redeemed

5. *THE MIGHTY MEN*

Like Modi of my T fame. May he note that T means option between two ways, one to the RH of God I Am another left – to the LAST JUDGMENT!

A. Modi may note that Siva was the prodigal son who left God I AM and Siva was operating through him in the recent days. His time is now over

B. Though the parable of the prodigal son does not give any time for return – Modi himself said that a time of 2 years is sufficient for appraisal. He said this in context of his ministers – but this also applies to him and HE FAILS

I tell you that man does not know of the kingdom of Bliss zones. And that the law of the new age is that happiness and bliss are given in proportion to what a man contributes or is poised to contribute for happiness of cosmos. I fought against Satanic spirits, I did not give them any happiness but I ensured happiness of cosmos of future and may be this is why God introduced the kingdom of bliss to me. Satanic spirits spied upon me and Sri Sri commented on G+ that there is a fountain of bliss in all human – but he is another idiot of Siva mafia and is on the blind precipice like many others

Indian Constitution starts with hypocrisy. It was made by a small mafia of Gandhi who started as Andhak the blind son of Siva and who tried to rape his mother Parvati. Later he became blind king Dhritrashtra = M K Gandhi a 3m1b Mahatma. If Modi and the blindly following world leaders do not see Truths of God – then no one can stop their fall from human kingdom forever?

Rev Book shows that Gandhi mafia used Antipas = Bhim Rao Am to make the Constitution of India and then spiritually murdered him. Worse they entered in gang conspiracy to garrote my Guaranteed Rights; even an UN Judge from Netherland joined these criminals. Do you think God is a fool?

How can they hide behind such Constitution which does not even belong to them? How can they hide behind slogan that they have full faith in Judiciary? How long will Indian fool their own souls?

Once again the Rev Book tells of the Depth of Satan. Siva and his mafia would enter anyone on top of governance or money control. First Congress, then BJP and also AAP, even other smaller parties

Now you know the truths but do not know how to overcome this. Much of the satanic mafia is dead but the human minds are on the same terrain and track and must overcome within – overcome their own ego, doubts, and static trends. Recently former PM M M

Singh who is the blue turban antichrist by Nostradamus gave a statement on overcome in a newspaper – I read on the title and knew that for him overcome means – to overcome Modi or Kejariwal. But this is not how God works. Even Sikh religion indicates that first one must overcome – dross within mind

6. *EVERY SLAVE*

This refers to men who were hypnotized or otherwise coerced by the enemy spirits. One can see this in standing ovations by US Congress to Narendra Modi whose right to enter USA was canceled by them. All men with mark of devil or the 666 mark on hands or forehead fall in this class. One indication of this is that large numbers of people blindly serve a guru even if he may be a criminal

- A. This trick and more can be read in Quran 113, 114
- B. Ash factor was in the foundations of the creation by 3nity. Ash given life by Siva made persons who were inclined to obey him, further forced to do so by hypnotism and many other tricks not worthy of the High God that he projected he was

7. *EVERY FREE MAN*

This refers only to the group of anti war activists led by Noam Chomsky and Susan Sarandon and this was foretold by God in Rev Book. I hope to find more especially amongst the wwYouth Intellectuals Judges and Journalists who have quest to know the Truths drive to set the earth right AND most importantly – for the passport to the new world orde

wwYouth

o Because - they belong to the next generation, even as the running generation of elders has by and large failed. Generation of Youth has onus to enter the new age and must RISE, if they follow the old trend they are liable to fall forever, but if they Rise they may intercede for their elders and save

Intellectuals

o Like Einstein and Noam Chomsky, though APJ Abdul Kalam failed himself to do Justice to my case. Noam Chomsky and Susan Sarandon led an anti war movement that God Hath foreseen and rewarded them there are others. Oppenheimer tested the first A-Bomb and compared the heat with light of a Gita verse. He was an intellectual who sold his soul to Satan. Soul of Einstein too was coerced by Satan. In India Kiran Bedi had the potential but her soul went to the side of Satan, first to Anna Hazare next to BJP. However I am open to save any and all who come with sincerity

Judges

- o   Let us face the Truth. Britain had the best judiciary and that without any written Constitution. Their first failure came after Justice Coke cornered the Crown King who back stabbed him. Another when the Privy Council did not indict Lord Clive in case filed by Nand Kumar, even a token punishment would have saved the credibility. As a rule when God gives something, He watches how you use this. Bible has a story about God giving a few coins to several persons and watching what they did with these. Britain not only failed to evolve Justice but also failed the BBC model. They failed to assist India, and they did the same and worse in USA which caused almost total fiasco of Will of God. Thus the sword and Arthur sank. Can Britain rise now is to be seen by them own?

I expect Judges to Rise and hope that CJI Thakur does this. On 15th August he said that he is at zenith of his career – but if he looks that way he gets lost, instead if he lays his life for Will of God he gets Eternal Life is clear message of Bible

He has the independent legal powers to have me address all Judges as in Isa 1.18, that would change the world and he would start on Way of Life...what use is zenith of career from where one falls in deep precipice, better is to start on Way of Eternal Life...I mean this CJI

Thakur and remember God would never come towards you, you have to take the first step towards Him and away from Modi. Of course CJI Thakur I tell you all this so that others may also listen and come -have a deal now! Never refuse a good offer – especially that which originates from God

Journalists

o Indian Journalists came to my side. But the media owners persecuted them. Later Siva took the Journalists away from me by his black magic

New Testaments has two references of the Journalists

1. Smyrna = a-z News
In this class those Journalists who serve the cause of God by spreading my Truths may get Crown of Life

2. eThyatira = wwNews
In this now that they know something about the Depth of Satan, they must come out of self imposed inhibitions and strive to get at the truths and to tell them to the masses. They may not be blocked by the media owners anymore because rule of Siva is over and media owners may go to eternal hell if they block the march of Truth. In this clause of Bible this is the burden

upon the Journalists and if they fail their head may be broken by rod of iron as if they are vessels of the potter

This book helps them to know of the depth of Satan, therefore I will make this mandatory for the Indian and Global national level Journalists to buy and read this book

8. *The Seven Churches*

The 7 churches of the Rev Book fall under class of the slaves – because the Spirit of God tells all them to Overcome such slavery of minds. Now that their captors have been finished it is easy for them to Overcome the black magical captivity of their minds. These churches are

A. Ephesus = Sheep US

- They refer to the Rev 6.9 persons of the US movement by Noam Chomsky and Susan Sarandon

B. Ephesus = a-z Sikhs

- The True Granth for them is Rev Book. Mary was their mother guru and with her 10 more disciples were the other gurus, a total of 11, the 12th Judas Iscariot = Aurobindo Ghose had failed in times of Jesus who too was not on the Right Hand of God but

Narayana spoke through him and set the pattern of the new age

- Another secret is that the Granth of Sikhs is attempting to imitate the directives of Quran 3.81. Gurtu Govind Singh gave directive of Granth but did not specify which Granth. My personal opinion is this that God I AM did not give precise information to Guru Govinda Singh to avoid more severe persecution of the Gurus and their kin

- There is one more secret. The disciples of Jesus did not evolve much. Only Mary Magdalene was granted immortality by the Lamb – Mary had a little lamb. The disciples were given ash factor treatment by Siva and God let them excel to see if Siva may change his attitude

- If this is true then the Gurus of Sikhs may RISE in the new age where they will be nurtured by God – let us be very clear that in present state the Sikhs do not have the quality to rule the world

C. Smyrna = S my name

- This is the name I have used on cover page of this book. My another divine name from Rev 3.12 is a2jnewj; a-z=Ahmad from Quran, another name is Smyrna = Ahmad4Allah; by Durga Saptsati my name is Manu Savarni who replaces the Vaivasvata Manu of Gita 4.2. My empirical name is Sudhir

- Smyrna = Name is A2z added with Rev 3.12 – I will write A2z New J on him add up as: Name is A2z, NewJ A2z in James Bond style

D. Smyrna = a-z Osama B L

- Is now not relevant on earth plane because he finished his tenure and made no successor

E. Smyrna = a-z News

- Is a special meaning and here the Journalists may get Crown of Life if they identify even an iota of their faith and endeavor to do Will of God I AM on earth as in the heaven

F. Pergamos = Buddhist G

- Buddha was an incarnation of Narayana, but not the final one. In most or all such cases Siva played the trick of remaking the apostle from ashes and in hope that if Maha Avatar comes in that person he will remain under Siva. That was the trap - now is the time that all branches of Buddhists can opt for the next step of evolution of humans

G. Thyatira = a2z Mahavira

- Similar to Buddhism and of the same time period – Mahavira came before Buddha but had a different role. Rev Book makes a special category of this church because eThyatira = wwNews. Times of India fulfills both the meanings. When I came with my concepts of God – TOI countered by a column named Speaking Tree. I countered that trees do not speak, far less deadwood trees and who spake were crows which are linked to Sita. Indu Jain is made from ashes of Sita and IS a queen of crows who crowed and crowed but of no advantage for mankind, then Modi took the baton from her. Had TOI stood for Truth they would have forced the Incompetent and Transgressor Judiciary of India to stop decay of the nation, but TOI opted to rather grow old like 175 not out but not wiser and nearer to God. Indu Jain also tried to tempt Krishna by a big poster on the front walls of the TOI building but God is not fond of gimmicks nor can Yoga ever be attained by technicalities not approved by God. Past Yoga gave some results because of mercy and patience of God but now the time is different. This is true that mahavira had some true Yoga but he walked with evil Chand Kaushik and fell

- By this Rev Book TRUTH, I place burden on those Journalists who now know something of the depth of Satan to promote my BOOK and Truth. I am now open to come before public and news media can spread truth in the public. If one soul is saved by anyone he too will be saved is a simple thumb rule. If more are saved there are higher rewards. If they fail to listen and act, their heads will be broken like iron rod breaks the vessels of the potter IS clear message to Thyatira church. But those who excel shall get the reward – to be a Morning Star. This is the opportunity for Indu Jain and whoso has courage for Truth of God I AM

H. Sardis = Parsi G

- In this I miss Bachi Karkaria, the enemy proved more powerful than her pull to my Truths. Now she can overcome. I tell you TOI has a real good team to be a pioneer of Truth which may liberate the earth. Indu Jain was once my mother – she can do more, but if not – who is my mother?

I. Sardis = Xians (1+1+1)

- Is for the Catholics and Protestants and Mormons. What I said to Indu Jain also apply to them. The enemy was far too powerful for the Vatican and other two Christian churches to have ever found the Truth

J. Sardis = M Mslm

- Quran 7.40 says that Muslims must believe Revelations of God, another translation says that they must believe signs of God. Either way – believe in Rev Book or the signs of Verbal Gospel algebra but without the door who I AM no Muslim goes to paradise is clearly written in Quran

k. Philadelphia = Siddha Yogi

- Refers to Ganespuri and New South Fallsberg establishment of Nityananada Muktananda Gurumayi. Rev Book clearly shows that

Philadelphia ka church = Synagogue e Satan

- Later I showed that all major Yoga systems of past fall under this class
  - o **Modi ka Yog = Philadelphia** is recent revelation
  - o **Philadelphia = Siddha Yogi**
  - o **Philadelphia = Sahaj Yog Ma** and
  - o **Philadelphia =- O! Integral** Yoga of **Aurobindo Ghose = Judas Iscariot** and **Ameeta Mehra = Babylons**

In context of fake and now bygone glory of India also note that Andhak the blind son of Siva became Dhritrashtra. Later he got a boon of having many eyes and became Dhritrashtra = M K Gandhi a 3m1b Mahatma. Here 3m means 3 monkeys and 1b the 1 bakari or goat

- Gandhi had gone to Africa to take charge of the Phoenix of Osiris the Satyavana whose wife Savitri was fooled by Yama who gave her a deceptive boon which delivered her in hands of 100 sons of Dhritrashtra. Cosmic management of Siva era was based on robbery of souls and such deceptions
- This very Yama was once Yudhister and delivered Draupadi in hands of those 100 sons. That time

Draupadi remembered Krishna who saved her. But as Savitri she did not remember Krishna and made blunder to trust Yama – the Vaivasvata Manu of Gita 4.2 whose clan now loses the Original Yoga

- One notes that Gandhi said – Be the change you want to see. What he saw was severe violence after the change done by him – that tells what he was. His life has more clues about his being Satan and he is father of India nation with throne of Satan
- He was enemy of God and his prayer indirectly blames God that He does not give Good Mind to all which causes all the problem of the world
- When I circulated a Vigilant Citizens paper with cartoon of CJI S S Mukerji, Justice V R Krishna Iyer said that Truth is not a defense against contempt of Court, I was frightened and escaped to Ganespuri Asram. Enemy knew that God I AM will give me Advanced Yoga and had conspired for long against plan of God
  - o They threw me out of job, brought to knees so that when I get Yoga from God they may hypnotize me to short sell That to them
  - o God gave me Big Gifts right within their Yoga establishment, they asked me the experience in writing hoping they can copy, I was unaware and did what they said. But they failed to copy

- o They tried all tricks, I was all the time unaware, I did not even know Who God IS, but God foiled all their tricks in mysterious ways
- o God led me to part ways. That was very difficult. Once a guru enters someone he grips him. But I threw them all off. One must pay Guru Dakshina to guru – I told God to pay them from my account

Siva being the ruler persecuted me. Recently I found that all this was written in Bible in context of a man who tortured his slave to repay the loan. Bible says that God told this man that when he begged from God He was soft unto him so how come he tortured me. In any case I did not owe anything to Siva except severe revenge for destroying Dwaraka 6 times and robbing my Yoga from my soul

Siva always had a powerful disciple to oppose incarnations of Narayana. Ravana opposed Ram. Krishna was opposed by many. This time he put Vanasura to oppose me. In times of Krishna there was a clear Word of God in Gita that either Arjuna must die fighting or win and rule. Here RULE meant to establish kingdom of God I AM on earth. Both Krishna and Arjuna forgot this and after the victory the RULE was given to Yudhister. That was the big blunder. Siva came as Durvasa and fooled Krishna to return. As He returned Yoga of Arjuna was robbed. This was done again and again as Dwaraka was made again and again and sunk by Siva 6 times to rob the Yoga – this time the table was reversed

- This is the main reason that Gita 4.3 says that Yoga of Ram clan was lost after a long time. The True Yoga is with me by Gita 4.3. Modi – this is your mirror if you can see

l. Laodiceans = One Lord

God I AM is the One God and by His Word I am the apostle of Deut 18. God I AM is the the One Lord God, the concept of One Lord was given in times of Moses and Christians and Muslims must also come to This Truth of the New Age because they are branches of the same old group

OVERCOME

Spirit of God says OVERCOME to all above 7 churches of the Rev Book. This means that they were under captivity of Siva, now that the influence is vanishing it is easier to overcome darkness of falsities and come to the side of the Light of Truth – this book is an aid for that. The time is ripe, the Siva team is either fully swept off or tied while waiting to avail mercy of God and they cannot block if you come to the side of God. Overcome also implies that you must show some strength to recognize and come to the side of Truth instead of worshipping statues and make believe in your good fortune based on blab la of Modi and Obama

As one example – whole world is going for statue of Gandhi. His concept was on self sufficient communes and never for mega cities. Why this hypocrisy? Answer is clear – a statue cannot find your faults, Chair is a statue of your goat, she cannot resist the daily rape. Ram Jethmalani is fond of judicial proverb – give a dog a bad name and hang it. New usage is – give a statue a good name and rape her. IS this worship of God, Modi?

RED ALERT
See page 154 for latest change in the cosmos

II. GOD MEANS BUSINESS – DEUT 18

Here is the essence of above prophecy by Moses

- *By Deut 18.15 – I am the Prophet of the prophecy of God by Moses. 18.16 – You cannot hear voice of God by your own choice made in the past, 18.17 – that is good for you, 18.18 – God Putteth His Words in My mouth, and I herein speak to you that God I AM Commandeth Me thus, 18.19 – God requires you to hear[1], 18.20 – If anyone does i me tation bandergiri about prophecy of God he dies, 18.21 – If you have any doubt if God Speaketh through Me, check thus, 18.22 – When I speak in the Name of God and that does not happen, you need not fear Me[2]*

- *Recently I wrote to SCI and several Indian Journalists about CHAOTIC weather – and the weather was so furious that I was shaken?*

- *My deeper studies show that God Himself does not involve on lower planes of I AM where the rules of those planes prevail. Those rules are based on Purusha and Prakriti. Purusha the Siva has gone and Prakriti the Virgin Mary has left Siva and is ready for – all things new*

- *This must be understood and severe destruction prevented. Modi leads all nations, so far he was doing this under hypnotism of Siva, now he may overcome and do what is wise – he may bring them to the Right Hand of God through me and limit himself to his jurisdiction*

- *I do not see any other way to save the nation or the world*

[1] This is stronger than the Shema speech by the Prophet Moses

[2] BY default – fear of God is starting point of wisdom for you

➤ 666 PUZZLE

About 18 years back I opened the first 666 puzzle of the Revelation Book – that was about Cain who is now Bill Clinton. After that several other persons were covered by 666 puzzles. India666 covered major persons implicated in my Case of Rights – who are destined to be maggots for millions of years, and that includes all Judges too – of SCI and HCs barring very few exceptions. In fullness of the prophecy very many others are covered. India666 named only the top men that if they may come back others too may be saved – but they are very shameless - to say the least

Today another 666 puzzle started to open. I was not inclined but God held me and waited till I found and solved the puzzle. This is hopefully the last puzzle and completes the cycle

This is about Revelation Chapter 13. This says that a second beast came up to work instead of wounded first beast. Also that those having mark of beast on hands (as in case of Sahaj Yoga of Nirmala Devi) or forehead (as in case of very many disciples of Ganespuri)[1] will survive and others die in The Judgment by Siva

Thus it is false that Siva wanted to save every person OR he wanted Antyodaya. What he wanted was always different than what he said – and that is what Modi IS!

The puzzle is:
666 = Q. Revelation Book Chapter 13 second beast e perdition?
A. Narendra Damodardas Modi

Obama and Modi are 2 horns of this Rev 13.11 beast

There are two beasts in Bible – one with 7 heads and 10 horns on who rides mother of all harlots. Another –

---

[1] And that includes Gandhi who once secretly visited Nityananda and did not write this in his biography – he had visited with a family friend Mahabala Swami and his family members told to a newspaper. Nityananda told Gandhi that India will not be free by fasting and eating bananas but he must launch Ram Vana. Gandhi thought and launched Bharat Chhoro movement. Vanguard of Gandhi was Lokmanya Tilaka a disciple of Siradi Sai who told him that now a bigger guru has come so Tilak may take things easy, further tasks will be done by another man. All events were well planned by Siva mafia

the SECOND beast of Rev 13.11 as above. These beasts and many others acted through many humans and were wounded healed, slain and remade and slain again – that is long story

Obama and Modi are 2 horns of this beast. Earlier pair was M M Singh the antichrist and George Bush. They are of Armageddon = Indo US. There are more pairs but are less important...Bible says that this beast showed magic of fire from sky, in case of this pair this refers to many events of lightning striking humans in India. That indicates that all similar events of mistreatment with mankind are handiworks of the Satan and not the idea of God...more is not within scope of this not that I know everything...

29 August 2016

❖ THEORY OF EVOLUTION

SPHINX

Sphinx and visions of Ezekiel and John Zebedee have four entities – a bull, lion, bird and human. These 4 tell the secret of human evolution

BULL

Bull is original symbol of Dharma and vehicle of Siva. The four aspects of Dharma are – Code of Ethics, Money, Desire and fulfillment, Liberation – that is back to Siva as ash

LION

Lion protected Parvati when she did severe taps to get Siva as her husband. Later the lion became vehicle of Durga made from portions of Deifies

In context of evolution the Lion means the fast change done by God through Virgin Mary and her child Lord Kalki

BIRD

In era of Siva the bird signified Garuda. There were two of these – one stillborn, second became vehicle of Visnu. One phoenix of Satyavana the Osiris was born from ashes and stolen by Gandhi and given to

Aurobindo Ghose = Judas Iscariot who named that as Supra Conscious

In context of the New Age the bird is male child of Rev 12.5 who became Lord Kalki

HUMAN

I am the Human on plane of earth. More is not necessary for the public to know at this stage

DR FREUD

He said that humans have essentially only two modes – work and love. This refers to above bull and lion. The third mode is esoteric – to evolve inwards like bird – and that is the way of becoming human. Siva failed in that and that was changed by God. In this a divine purpose was served that the history of past reminds of the tree of knowledge of good and bad. Now those who come to the Right Hand of God will eventually eat of the tree of Life and then they would rule the cosmos – having known good and bad they would be superior administrators. The present rule that power of muscle or money can make rulers is now not valid and this applies to power of muscle of tongue of Modi too

HUMAN EVOLUTION

DARWIN

Darwin told of theory of adaptation, survival by fight or flight. This theory applied to animals who belonged to level 4 of I AM as reckoned from below. The clear meaning is - and Darwin did not know this – that the adaptation is by that quantum I AM

Fight or flight comes from a brain part reticulam that comes from reptiles and this gives snake type qualities of attack or running away

Level 3 I AM is in plants and reflects as response to gravity and sun and several other qualities. The Level 2 I AM reflects in crystals and most specially in ice crystals. Level 1 I AM is of infra matter and much is known about that plane of reality

HUMAN

Next to animals come humans. amygdala develops to substitute reticulum. By this humans can insist of higher truths and also fight Dharma Yuddha. But this brain organ is not sufficiently evolved in humans and they behave as yet like animals when it comes to survival. This happened mostly because Siva was prodigal son and left God. He was always worried about survival and the natural human evolution was stunted and even reversed – man becoming worse than animal

Genius

At this level the good civilizations must emerge – both fight and flight responses of animals must come to higher and civilized levels. For the reason told above the matters became worse – the wisdom is used for TRICKS, and now the animals have most fierce fangs and claws even atomic swords and spears. In context of the new age the following provisions apply to those wise ones who are perverted – one example is the person who exploded atom bomb and read verse of Gita that the light of that explosion was like millions of Suns

Prophet

They were all made by Siva and now all shalt either repent and return or perish. Modi is trying to unite and reorganize of what is left of all spiritual prophets of the past era and the only Way of Life for him is to repent and return asap

God Man

This is the Level of Chosen One – my level. Kalki of John 3.16 is said to be the Only begotten Son – of man having the mysteries of God, therefore one conjectures that others have to follow me and there shall not be another like me

GOLD OF GOD

The Alchemic Gold of God can be given only by God and through Kalki and me – there is no other way of human evolution, therefore the mankind needs to follow me

PROMISED LAND

Moses was not given the Promised Land but was shown. Siva could not get his ashes. Both these indicate that Moses shall be reborn in new age. This is false propaganda of Siva mafia that followers of Moses entered the Promised Land. How can Moses be omitted?

Promised Land has been given to me. No one else gets that

THEREFORE FOLLOW ME!

ABOUT ME

My one divine name is A2znewJ[1]. I AM that apostle of the God of who Moses told in Deut 18. I am 69 years of age and an engineer by education. In this life my journey on the Way started on Xmas 1974 like this

My cash got lost in Rome and I found myself weeping before La Pieta in Vatican unaware that she was my ancestor mother Mary Magdalene

I was hungry, food was the first priority. I had with me air ticket of Paris and decided to proceed so that I get food in the plane. But the Air India plane was hijacked and circled on Rome. A thought came that why I am worried merely for food while hundreds of humans may crash if the fuel is finished. Eventually the plane landed but did not take any passengers from Rome – NO FOOD FOR ME. Other airlines would not take me because my ticket was bought with youth concession...

From here started the miracles of God I AM. Later I learnt that God I AM was working on me since I was a child named Dhruva – about 80000 years back

---

[1] By Rev 3.12 puzzle – I will write = a2znewj on him

ABOUT THIS BOOK

I wrote this in a very short time – all through every day or couple of days God was revealing a new truth to me. That was inserted in most appropriate running context of the text already written, and a date was put for my own memorandum. Having been written in this way the dates are not in chronological sequence but put according to context

Yesterday the 20 August 2016 an Indian TV channel showed a research on Radha. I had done my own research in Vrindavana but missed just one place – the Mansarovara where Radha came after Siva transgressed the Ras Lila in disguise of a woman. This story is told in many ways but the TV channel got the correct version. This further shows that not only Siva fails but he was controlling and fooling even Krishna who could not deal with strict hand, therefore he too fails but Radha is redeemed. This is similar to Visnu having failed but Tulasi being redeemed. Further investigations of 30 August 16 showed that Krishna was always with God and when he seemed on a different track that was misleading by Siva. Krishna will soon be incarnating on earth

On 4 Sept 2016 God I AM announced the Biblical Flood – spiritual flood to last for 12 years and I included that as Breaking News, also deleted about 100 pages of this book which became less signifacnt

GOD I AM – THY WILL BE DONE, THY KINGDOM COME, ON THE EARTH AS IN THE HEAVEN

1 September 2016 - Swasti

On Sept 4 I was told by God I AM to tell of the decision to make all things New – I cut short my book from 276 to now 164 pages and hurried to get it published – for earliest savings. The most sinful issues on which Siva was tested include my case of Rights and sins against Holy Spirit the details of which are not known to you. Modi was taken hold of by Siva and Obama by Ganesa and they also made fresh Indo US = Armageddon. Other nations and UN were also hypnotized even as UN was earlier hypnotized by Gurumayi too. I also learnt that problems of weather and quakes as also of terrorism are being caused by Siva – as depth of Satan and God I AM had been neutral. This is implied that those who do not make distance with Modi may not be eligible for saving. My counsel is that everyone carries his own cross and do not mix up the matters

BREAKING NEWS

Biblical Flood started on Sept 3, 2016

Next day the Sept 4, 2016 the Biblical prophecy of curse on gold silver precious stones and money was open

NATURE OF FLOOD

This is opposite to the theory of creation in Sankhya Yoga – and is also described in Uddhava Gita. I do not know if that version is technically correct or the text is distorted with time – but I know that when Narayana blessed me as Dhruva then He gave me Gnosis to sing His Prayer and the last verse tells of dissolution of creation back in MAHAT that is The Cosmic Mind. In this life I have ben on toes and did not get time to read and understand that Prayer

DURATION OF THE FLOOD

This is 12 years and will be selective

KINGDOM OF VANASURA

Immediately before the Breaking News the kingdom of Siva Ganesa and Vanasura were fuly wiped off from all spiritual planes. I do not know what effect it will have on the physical plane, but that is also of lesser relevance

SAVING

In my role as Savior – I invite the Kings of the nations with Judges and Ministers – as in Isa 1.18 so that they can be made an offer from God and have option to make a covenant as told in Bible and also Quran

For this they have to pay $500 on my site apostleofthegod.wixsite.com/flood

They will be entitled to receive guidelines on how to proceed further – at their own expenses an offerings will apply. They have to intercede for their nation and go by the guidelines from time to time. They will be required to make offerings for them their nations some for pool and some to the apostle

THE LAST DATES

For India the last date is 30 September 2016 and for other nations 31 December 2016

RICH MEN

They can apply independently in the same way if they so like – there are no last dates for them

TERMS APPLY

Most important one is that elder generation will be given Yoga only in next birth after tare and wheat is separated. In this life only youth will be given Yoga and can intercede for elders for full assurance of their saving

MOTHER TERESA

She was granted status of Saint – yesterday in Vatican. This church is church of 3nity=Babylon and has no jurisdiction. My journey that started from Vatican seems to complete the cycle by this comment on this False Church – Mother Teresa herself would rather have saved the money spent on this function and spent that to help flood victims of India. The church must surrender or fall and come to the Right Hand of God

Mr Pope, by Urantia Book the Vatican has corrupted the Rev Book and for that sin alone they get their proper wages – you know what, therefore surrender lest you fall for ever in animal zone

1.16 PM 5 Sept 2016 – End of this Book

Contents